THERMOPLASTIC MATERIAL SELECTION

PLASTICS DESIGN LIBRARY (PDL)
PDL HANDBOOK SERIES
Series Editor: Sina Ebnesajjad, PhD (sina@FluoroConsultants.com)
President, FluoroConsultants Group, LLC
Chadds Ford, PA, USA
www.FluoroConsultants.com

The **PDL Handbook Series** is aimed at a wide range of engineers and other professionals working in the plastics industry, and related sectors using plastics and adhesives.

PDL is a series of data books, reference works and practical guides covering plastics engineering, applications, processing, and manufacturing, and applied aspects of polymer science, elastomers and adhesives.

Recent titles in the series

Biopolymers: Processing and Products, Michael Niaounakis (ISBN: 9780323266987)
Biopolymers: Reuse, Recycling, and Disposal, Michael Niaounakis (ISBN: 9781455731459)
Carbon Nanotube Reinforced Composites, Marcio Loos (ISBN: 9781455731954)
Extrusion, 2e, John Wagner & Eldridge Mount (ISBN: 9781437734812)
Fluoroplastics, Volume 1, 2e, Sina Ebnesajjad (ISBN: 9781455731992)
Handbook of Biopolymers and Biodegradable Plastics, Sina Ebnesajjad (ISBN: 9781455728343)
Handbook of Molded Part Shrinkage and Warpage, Jerry Fischer (ISBN: 9781455725977)
Handbook of Polymer Applications in Medicine and Medical Devices, Kayvon Modjarrad & Sina Ebnesajjad (ISBN: 9780323228053)
Handbook of Thermoplastic Elastomers, Jiri G Drobny (ISBN: 9780323221368)
Handbook of Thermoset Plastics, 2e, Hanna Dodiuk & Sidney Goodman (ISBN: 9781455731077)
High Performance Polymers, 2e, Johannes Karl Fink (ISBN: 9780323312226)
Introduction to Fluoropolymers, Sina Ebnesajjad (ISBN: 9781455774425)
Ionizing Radiation and Polymers, Jiri G Drobny (ISBN: 9781455778812)
Manufacturing Flexible Packaging, Thomas Dunn (ISBN: 9780323264365)
Plastic Films in Food Packaging, Sina Ebnesajjad (ISBN: 9781455731121)
Plastics in Medical Devices, 2e, Vinny Sastri (ISBN: 9781455732012)
Polylactic Acid, Rahmat et al. (ISBN: 9781437744590)
Polyvinyl Fluoride, Sina Ebnesajjad (ISBN: 9781455778850)
Reactive Polymers, 2e, Johannes Karl Fink (ISBN: 9781455731497)
The Effect of Creep and Other Time Related Factors on Plastics and Elastomers, 3e, Laurence McKeen (ISBN: 9780323353137)
The Effect of Long Term Thermal Exposure on Plastics and Elastomers, Laurence McKeen (ISBN: 9780323221085)
The Effect of Sterilization on Plastics and Elastomers, 3e, Laurence McKeen (ISBN: 9781455725984)
The Effect of Temperature and Other Factors on Plastics and Elastomers, 3e, Laurence McKeen (ISBN: 9780323310161)
The Effect of UV Light and Weather on Plastics and Elastomers, 3e, Laurence McKeen (ISBN: 9781455728510)
Thermoforming of Single and Multilayer Laminates, Ali Ashter (ISBN: 9781455731725)
Thermoplastics and Thermoplastic Composites, 2e, Michel Biron (ISBN: 9781455778980)
Thermosets and Composites, 2e, Michel Biron (ISBN: 9781455731244)

To submit a new book proposal for the series, or place an order, please contact David Jackson, Acquisitions Editor
david.jackson@elsevier.com

THERMOPLASTIC MATERIAL SELECTION
A Practical Guide

Eric R. Larson

Amsterdam • Boston • Heidelberg • London
New York • Oxford • Paris • San Diego
San Francisco • Singapore • Sydney • Tokyo

William Andrew is an imprint of Elsevier

William Andrew is an imprint of Elsevier
The Boulevard, Langford Lane, Kidlington, Oxford, OX5 1GB, UK
225 Wyman Street, Waltham, MA 02451, USA

Copyright © 2015 Elsevier Inc. All rights reserved.

No part of this publication may be reproduced or transmitted in any form or by any means, electronic or mechanical, including photocopying, recording, or any information storage and retrieval system, without permission in writing from the publisher. Details on how to seek permission, further information about the Publisher's permissions policies and our arrangements with organizations such as the Copyright Clearance Center and the Copyright Licensing Agency, can be found at our website: www.elsevier.com/permissions.

This book and the individual contributions contained in it are protected under copyright by the Publisher (other than as may be noted herein).

Notices
Knowledge and best practice in this field are constantly changing. As new research and experience broaden our understanding, changes in research methods, professional practices, or medical treatment may become necessary.

Practitioners and researchers must always rely on their own experience and knowledge in evaluating and using any information, methods, compounds, or experiments described herein. In using such information or methods they should be mindful of their own safety and the safety of others, including parties for whom they have a professional responsibility.

To the fullest extent of the law, neither the Publisher nor the authors, contributors, or editors, assume any liability for any injury and/or damage to persons or property as a matter of products liability, negligence or otherwise, or from any use or operation of any methods, products, instructions, or ideas contained in the material herein.

British Library Cataloguing-in-Publication Data
A catalogue record for this book is available from the British Library

Library of Congress Cataloging-in-Publication Data
A catalog record for this book is available from the Library of Congress

ISBN: 978-0-323-31299-8

For information on all William Andrew publications
visit our website at http://store.elsevier.com/

Publisher: Matthew Deans
Acquisition Editor: David Jackson
Editorial Project Manager: Peter Gane
Production Project Manager: Nicky Carter
Designer: Mark Rogers

Typeset by TNQ Books and Journals
www.tnq.co.in

Printed and bound in the United States of America

To all those who dare to dream

Acknowledgments

I would like to thank the members of my graphics team for helping create a number of images that were used in this book. Whether it was photography, illustration, digital editing, or the thankless task of organization, you guys were amazing.

There is an old saying: A picture is worth a thousand words. All I can say is that your contributions were appreciated. Thank you.

Alex Rennie
Dream Catcher Creative Media
http://dreamcatchercreativemedia.com/

Monique Feil
Monique Feil Photography
http://moniquefeil.com

Zach Petschek
Zach Petschek Photography
http://zachpetschek.com/

Contents

Preface xv

1 Introduction 1
   1.1   The Stone Age 1
   1.2   The Age of Metals 3
   1.3   Other Materials 7
   1.4   The Industrial Revolution 8
   1.5   Mass Production 10
   1.6   Materials Science 11
   1.7   The Plastics Age 12
   1.8   Plastics—The Other Synthetic Material 14
   1.9   Plastics Material Selection 15
   1.10 How This Book Can Help You 16
   References 17

2 Why Use Plastic? 19
   2.1   Introduction 19
   2.2   Plastics as Raw Materials 20
        2.2.1   Thermosets 21
        2.2.2   Thermoplastics 21
        2.2.3   Key Differences 22
   2.3   Plastic Processing Technologies 24
        2.3.1   Plastics Tooling 25
        2.3.2   Plastics Forming 26
        2.3.3   Vacuum Forming 26
        2.3.4   Pressure Forming 27
        2.3.5   Plastic Welding 28
        2.3.6   Extrusion 29
        2.3.7   Plastics Molding 30
        2.3.8   Casting 30
        2.3.9   Compression Molding 31
        2.3.10 Rotational Molding 32
        2.3.11 Injection Molding 32
        2.3.12 Reaction Injection Molding 34

		2.3.13	Transfer Molding 34

- 2.3.13 Transfer Molding 34
- 2.3.14 Structural Foam Molding 35
- 2.3.15 Expandable Foam Molding 36
- 2.3.16 Blow Molding 37
- 2.3.17 Extrusion Blow Molding 37
- 2.3.18 Injection Blow Molding 38
- 2.4 Process Comparison 39
- 2.5 Plastics in Manufacturing 42
  - 2.5.1 Mass Production 42
  - 2.5.2 Batch Production 43
  - 2.5.3 Job Production 44
  - 2.5.4 Manufacturing Processes 44
- 2.6 Advantages of Thermoplastics 46
  - 2.6.1 Performance 46
  - 2.6.2 Processing Options 47
  - 2.6.3 Near-Net-Shape Manufacturing 47
  - 2.6.4 Safety 48
  - 2.6.5 Cost 48
- 2.7 Disadvantages of Thermoplastics 49
  - 2.7.1 Heat Resistance 49
  - 2.7.2 Time-Dependent Behavior 49
  - 2.7.3 Temperature Variations 51
  - 2.7.4 Structural Inconsistencies 51
  - 2.7.5 Repairing Broken Parts 52
  - 2.7.6 Perception 53
  - 2.7.7 Human Behavior 54
- 2.8 The Uniqueness of Thermoplastics 55
- 2.9 And the Answer Is… 56
- References 56

3 Understanding Thermoplastics 57
- 3.1 Introduction 57
- 3.2 Materials Science 58
  - 3.2.1 Strength of Materials 58
  - 3.2.2 Anisotropic Behavior 60
  - 3.2.3 Nonlinear Behavior 62
  - 3.2.4 Stiffness 62
  - 3.2.5 Toughness 63
- 3.3 Polymer Science 64

- 3.4 The Resin Industry 69
  - 3.4.1 Resin Production 70
  - 3.4.2 Resin Distribution 71
  - 3.4.3 Resin Grades 75
  - 3.4.4 Resin Modification 79
  - 3.4.5 Processing Aids 81
  - 3.4.6 Fillers 82
  - 3.4.7 Reinforcements 82
  - 3.4.8 Performance Modifiers 83
  - 3.4.9 Resin Versions 84
  - 3.4.10 Alloys and Blends 86
  - 3.4.11 So Where Is the Data? 87
- 3.5 Thermoplastic Classification Methods 90
  - 3.5.1 Amorphous versus Semicrystalline 92
  - 3.5.2 Chemical Family 93
  - 3.5.3 Cost versus Performance 93
  - 3.5.4 Elasticity 94
- 3.6 A Final Word about Property Data 94
- 3.7 The Amazing World of Thermoplastics 94
- Further Reading 95

4 An Overview of Thermoplastic Materials 97
- 4.1 Key Thermoplastic Materials 97
  - 4.1.1 Commodity Plastics 98
  - 4.1.2 Engineering Plastics 108
  - 4.1.3 Specialty Plastics 122
- 4.2 Thermoplastic Elastomers 135
- 4.3 Meet the Family 141
- Further Reading 143

5 Material Selection Based on Performance 145
- 5.1 What is Performance? 145
- 5.2 Predicting Performance 147
  - 5.2.1 Correlation 148
  - 5.2.2 Wrong Criteria 149
  - 5.2.3 Disruptive Innovation 149
- 5.3 How Material Selection Affects Performance 150
  - 5.3.1 Evaluating Property Data 152
  - 5.3.2 The Importance of Design 152

	5.3.3	The Importance of Processing 154
	5.3.4	Property Data—A Final Caveat 155
5.4	Environmental Effects 156	
	5.4.1	Temperature 158
	5.4.2	Chemicals 159
	5.4.3	Radiation 160
	5.4.4	Time 162
5.5	Key Mechanical Properties 164	
	5.5.1	Strength 164
	5.5.2	Stiffness 166
	5.5.3	Toughness 167
5.6	Measuring Toughness 170	
	5.6.1	Izod Test 170
	5.6.2	Un-Notched Izod Test 171
	5.6.3	Charpy Test 171
	5.6.4	Gardner Impact Testing 172
	5.6.5	Instrumented Impact Tests 173
	5.6.6	High-Speed Tensile Tests 173
	5.6.7	Projectile Testing 174
	5.6.8	Drop Testing 174
	5.6.9	Tumble Testing 175
5.7	But Is It Tough Enough? 176	
	5.7.1	Are You Ready to Rumble? 180
	5.7.2	Cutting the Grass 181
	5.7.3	Chicago Style 182
	5.7.4	The Thrill of Victory (and the Agony of Defeat) 184
	5.7.5	Dealing with JRA 185
	5.7.6	The Shupe Test 187
	5.7.7	The Bottom Line on Toughness 188
5.8	Surface Properties 188	
	5.8.1	Friction 189
	5.8.2	Lubricity 189
	5.8.3	Wear 190
	5.8.4	Hardness 190
5.9	Key Electrical Properties 191	
	5.9.1	Insulating Plastics 191
	5.9.2	Conductive Plastics 191

Contents

- 5.10 Properties of Form 192
  - 5.10.1 Size 192
  - 5.10.2 Shape 194
  - 5.10.3 Appearance 195
- 5.11 Some Final Guidelines 197
  - 5.11.1 Conceptual Tools 198
  - 5.11.2 Mathematical Tools 199
  - 5.11.3 Determining Critical Material Properties 203
- References 204
- Further Reading 205

6 Material Selection Based on Cost 207
- 6.1 What is Cost? 207
- 6.2 Why is Cost Important? 209
  - 6.2.1 The Way We Measure Things 210
  - 6.2.2 Relationship to Performance 210
  - 6.2.3 Business Perspective 210
  - 6.2.4 The Bottom Line on Cost 211
- 6.3 The Language of Cost 212
- 6.4 Evaluating Cost 214
  - 6.4.1 Material Cost 215
  - 6.4.2 Processing Cost 216
  - 6.4.3 Process Rates 216
  - 6.4.4 Cycle Times 217
  - 6.4.5 Adding Up the Numbers 219
- 6.5 Reducing Material Costs 219
  - 6.5.1 Optimize the Structure 220
  - 6.5.2 Effective Specifications 222
  - 6.5.3 Optimize the Wall Thickness 224
  - 6.5.4 Exploit the Material 224
  - 6.5.5 Exploit Competitive Advantages 227
  - 6.5.6 Saving Pennies 230
- 6.6 Reducing Processing Costs 230
  - 6.6.1 Optimize the Geometry 231
  - 6.6.2 Effective Specifications 233
  - 6.6.3 Uniform Wall Thickness 238
  - 6.6.4 Exploit the Material 239
  - 6.6.5 Design for Speed 241
  - 6.6.6 Integrating Team Input 242

6.7	Total Manufacturing Cost		243
	6.7.1	Explore Design Options	243
	6.7.2	Do the Math	243
	6.7.3	Find an Edge	247
6.8	A Final Word on Cost		247
References			250

7 Material Selection Based on Feel ... 251

7.1	What Is Feel?		251
7.2	Why Is Feel Important?		253
	7.2.1	Product Performance	253
	7.2.2	Sales and Market Share	254
	7.2.3	Technical Validity	255
	7.2.4	The Bottom Line on Feel	256
7.3	The Language of Feel		256
	7.3.1	Sensory Input	257
	7.3.2	Human Response	259
	7.3.3	Not an Engineering Language	263
	7.3.4	An Imprecise Language	264
	7.3.5	Comparative Analysis	265
7.4	Evaluating Feel		265
	7.4.1	Physical Equipment	266
	7.4.2	Human Senses	266
7.5	Sight		266
	7.5.1	Light	267
	7.5.2	Color	269
	7.5.3	Patterns	270
	7.5.4	Material Selection Based on Sight	271
	7.5.5	Opportunities	273
7.6	Hearing		273
	7.6.1	Sound	273
	7.6.2	Vibration	274
	7.6.3	Acoustics	274
	7.6.4	Psychoacoustics	275
	7.6.5	Music	276
	7.6.6	Human Response	277
	7.6.7	Material Selection Based on Hearing	277
	7.6.8	Opportunities	279

7.7 Touch 279
    7.7.1 Size and Shape 280
    7.7.2 Weight and Density 280
    7.7.3 Temperature 282
    7.7.4 Pressure 286
    7.7.5 Vibration 286
    7.7.6 Movement 287
    7.7.7 Hardness 289
    7.7.8 Texture 291
    7.7.9 Slipperiness 291
    7.7.10 Material Selection Based on Touch 292
    7.7.11 Opportunities 293
7.8 Smell 293
    7.8.1 Odor Detection 294
    7.8.2 Odor in Thermoplastics 296
    7.8.3 Human Response to Odor 298
    7.8.4 Material Selection Based on Smell 299
    7.8.5 Opportunities 299
7.9 Taste 300
    7.9.1 Material Selection Based on Taste 301
    7.9.2 Opportunities 302
7.10 A Methodology 302
    7.10.1 Infrastructure 303
    7.10.2 The Process 304
    7.10.3 Making It Work 306
7.11 A Final Word about Feel 308
References 308
Further Reading 310

8 Bringing It All Together 311
8.1 Material Selection 311
    8.1.1 Step One: Establish Key Criteria 312
    8.1.2 Step Two: Select the Process 312
    8.1.3 Step Three: Develop a Short List 312
    8.1.4 Step Four: Evaluate the Data 312
    8.1.5 Step Five: Development 313
    8.1.6 Step Six: Select the Material 313

8.2	Material Specification		313
	8.2.1	Approved Suppliers	314
8.3	The Plastics Supply Chain		315
	8.3.1	Resin Suppliers	316
	8.3.2	Compounders	316
	8.3.3	Converters	316
	8.3.4	Equipment Suppliers	316
	8.3.5	Toolmakers	317
	8.3.6	Product Manufacturers	317
8.4	Industry Infrastructure		317
	8.4.1	Trade Organizations	318
	8.4.2	Education	318
	8.4.3	Information Providers	319
	8.4.4	Plastics Testing	319
	8.4.5	Service Providers	319
8.5	Working with Suppliers		320
	8.5.1	Determining Capabilities	320
	8.5.2	Determining the Right Fit	321
	8.5.3	Project Participation	321
	8.5.4	Managing the Relationship	322
	8.5.5	Communication	322
8.6	Troubleshooting		322
	8.6.1	Assemble the Team	323
	8.6.2	Identify the Real Problem	323
	8.6.3	Determine the Origin	324
	8.6.4	Understand the Root Cause	327
	8.6.5	Solve the Problem(s)	328
8.7	Finale		328

Resources 331
Index 339

Preface

I have been building things for as long as I can remember. It started when I was a kid with a set of Tinker Toys. With a floor to play on and a tube full of sticks and wheels and paddles, I could let my imagination run wild. I would spend hours making anything and everything. Then there were Lincoln Logs and Lego blocks and Erector Sets. And while it was fun to build things with those kits, Tinker Toys were always my favorite.

When I got older, I started building tree houses and forts, then model cars and model airplanes, toothpick bridges, you name it. I loved exploring how things went together, and how they came apart. Sometimes I would take something apart just to see how it was put together. Then, I would rebuild it—some times with "improvements"—other times with things exactly the same. Although, more often than I would like to admit, there were times when I thought I had put something back together exactly the way I had found it, only to discover that I had overlooked a part, which was now lying on the workbench. Oops!

As I experimented with building things, I would often look at the things that were around me, not just the simple things, but big important stuff like houses and cars and bridges, even airplanes. I dreamed that one day I could help build something big and important like that. I did not really think about what they would be made of—wood, concrete, steel, whatever—I just knew that I wanted to help build them. When I got to college, I ended up studying aerospace engineering. My goal was to build airplanes.

After I got my degree, I got a job working for an aircraft company. One of my first assignments involved the design of the support structure for a vertical stabilizer, what is commonly known as a rudder. It is an important part of the aircraft, and if it does not work, well, there are going to be problems. I do not remember much about the job, except for the fact that I hated it. I used to try and describe my job to my friends. I would show them pictures, and point to the parts I had worked on. *Hey, look at this structural support, and the fasteners that were used. No, not that one, the over one, the third one from the bottom, the one with no rivets, just a single machine screw with a locking washer. That one. Isn't that an awesome design?* Needless to say, I soon realized that aircraft design was not my true calling.

So, I quit my job and got a job designing body boards. A body board is a short surfboard, usually made of foam, that allows the user to lay on top of it, and then ride the surf in a manner similar to body surfing. While the job did not pay that well, it was a fun job, and every thing about it was interesting. I learned about design, materials, manufacturing costs, product development, and listening to the voice of the customer. It was also my introduction to the world of thermoplastics.

In the time since then I have had a number of different jobs, most of them in the world of product design and development, and in all of them I have worked with thermoplastics—either as a design engineer, application engineer, or project manager. In the process, I have learned a thing or two about these materials, and I have tried to capture some of the things I have learned in this book. My intent is not to present a comprehensive encyclopedia about thermoplastic materials or to pretend that I know everything there is to know about these materials. Rather, I am presenting a guide book, one that provides practical, common sense information about the use, reuse, and effective application of these unique materials.

My hope is that those who read this book will use the information in a wise and thoughtful manner—not just to make something cheaper, but to devise new and better ways of making things, things that will change our world for the better. In the words of Walt Disney, *If you can dream it, you can do it.*

Eric R. Larson
January, 2015

1 Introduction

Whether tools, weapons, clothing, shelter, jewelry, or even toys, humans have always made things. One might even argue that the meaning of the word human is *to make*.

The materials used to make things have changed throughout history, from found to extracted to synthesized. In the process, knowledge about how to use these materials has been passed along from generation to generation, from culture to culture, from continent to continent. This knowledge comes in a variety of forms, ranging from common sense, to tribal knowledge, to sage wisdom, to trade secrets and intellectual property.

Today, many of the things that humans make are made from plastics. We use plastics to protect our children, to preserve our food, to entertain us, and to communicate and connect with other people. Our health and happiness in our day-to-day lives—and, one might even argue, the very future of our species—depends on our effective use of these materials.

Plastics are unique materials, with unusual properties and performance characteristics. They range from simple compounds similar to beeswax, to highly engineered, specialty materials like Teflon®, whose properties seem to defy the laws of physics. After all, if Teflon is such a resilient, repellant, and nonstick material—how does it stick to the pan?

There are many books about plastics technology, including some that try to help users select an appropriate plastic material for a given application. Sadly, most of these books are written from the perspective of a polymer chemist, and they fail to provide the user with any guidance on how to evaluate plastic materials in a practical, hands-on manner.

This book is meant to be a guide in the process of plastics material selection. It is based on the simple premise that we all make things, and that we have a fundamental understanding of how to use materials—based on our heritage as human beings.

Let us begin our journey into the world of plastics by taking a quick look at human history. You may find that you know more about materials—and about plastics—than you think.

1.1 The Stone Age

Paleoanthropology—the study of ancient hominid fossils—has shown that humans have always made things. The oldest stone tools, found in

Ethiopia, date back to about 3.4 million years ago, and could have been used by any of the several varieties of hominid, perhaps even by the ancestors of our species, *Homo sapiens* [1]. Less than a million years ago, Neanderthals, a not-so-distant cousin (and perhaps a subspecies of *H. sapiens*), left behind musical instruments made out of bones, along with stone tools and flint blades. They used these blades to make wooden spears, hand-axes, and to skin animal hides for clothing and shelter. Historical records of our species, *H. sapiens*, date back to about 200,000 years ago. It is now known that *H. sapiens* interacted with Neanderthals, resulting in some genetic exchange, and probably some material culture sharing, too [2–4] (Figure 1.1).

Regardless of the species, all of the earliest hominids used naturally occurring materials, such as plants, bones, feathers, skins, and tendons. Humans found and modified shells, antlers, and horns for functional and decorative purposes. Shells coated in red ochre clay and used as beads for jewelry, found in what is now eastern Morocco, date back to 82,000 years ago. Excavations in the Sibudu Cave alongside the Tongati River in South Africa reveal shells that were used as containers—not just for water or other goods, but for paint (see Ref. [5]). Layers of sediment preserving

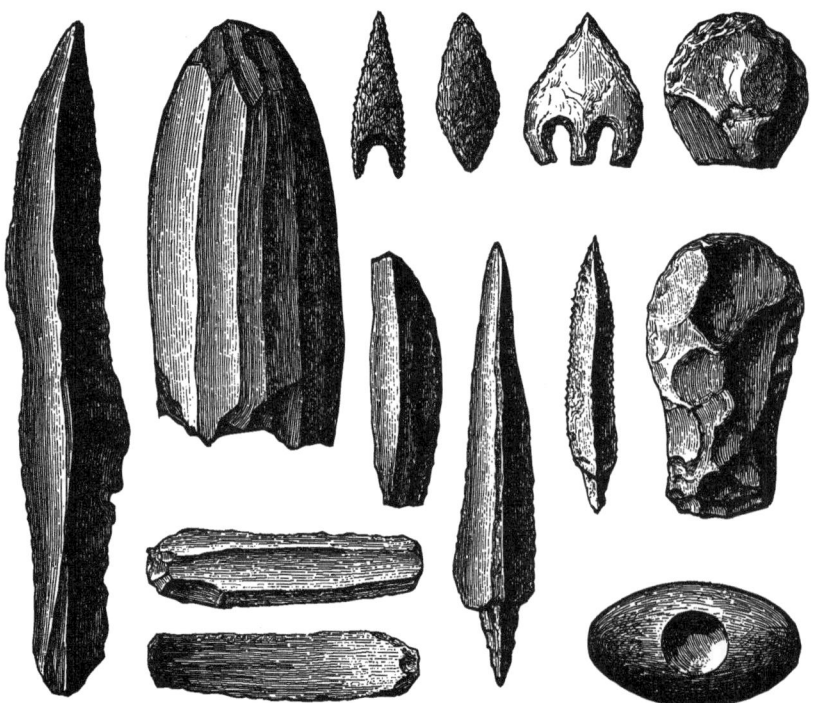

Figure 1.1 A sample of early stone tools. HeinNouwens/Shutterstock.com.

years of occupation dating from 50,000 to 80,000 years ago contain needles, animal traps, bone arrowheads, some stone tools, and a type of binder or glue made of ochre clay [6]. Much later, stone was used to make massive buildings and, around the time the wheel was invented, the first roads.

The materials these early humans left behind have enabled anthropologists to learn about each civilization's technological and cultural development. The things they made and how they made them—their material culture—can tell us a lot about how they lived, what was important to them, and what motivated them to build and invent new things. Much of the knowledge about materials came from the exchange of ideas and material culture between peoples, as well as apprenticeship, the passing on of knowledge through generations. As it turns out, there was a "materials science" long before the term was ever coined and institutionalized in universities.

As humans developed, so did their use of materials. Many of the materials used in ancient times are still in use today, including stone and animal shells. These materials still have the same qualities that they did back then, including appearance and feel. But the development of metals led to a whole new way of life, made possible by the unique qualities of these materials. While it was once thought that the practice of metallurgy began in one place and diffused around the globe through trade, the latest archaeological evidence indicates that it was more complicated than that. In many cases, people discovered how to use metal independently of one another. This led to a diversity of techniques in metallurgy, some of which led to smelting, as well as the synthesis of new alloys.

1.2 The Age of Metals

Beginning about 8700 BCE, copper with its lustrous sheen and flexible properties attracted the human eye in Mesopotamia, Asia and other sites around the world. People learned how to extract raw metal from ores, and began to use found metals like copper, silver, and gold. Europeans and Asians first used copper to make pigments and jewelry [7]. Smelted copper appeared around 5500 BCE in southeastern Europe—what is now Serbia [8]. The Sumerians and Egyptians also loved copper, crafting religious iconography, jewelry (of course), and even pipes for use in plumbing [9,10]. Copper was used in North and South America, too. Some of the oldest copper artifacts date back to about 6000 BCE in North America. Indigenous Americans did not smelt copper, but hammered it into shapes. Living near the Great Lakes, they had access to a great deal of pure copper, and made knives, awls, spearpoints, and jewelry [11,12] (Figure 1.2).

Figure 1.2 A sample of copper ore. RussellShively/Shutterstock.com.

With the advent of smelting, metals could not only be extracted and purified, but also intentionally combined with other materials while in a molten state. The resulting metal alloy would then be poured into a cast, usually made of clay, and cooled until hardened in the desired shape. Bronze, the first alloy, was primarily made of copper combined with smaller amounts of arsenic or tin. It first appeared as a copper and tin mixture in the Near East, Egypt, and Mesopotamia. The oldest known artifacts were produced by the relatively peaceful Vinča farming culture in what is now Serbia, and included figurines and ornaments [13]. Around the world, bronze became the primary metal for use in toolmaking, as it was much stronger than copper. In China, bronze remained the preferred metal for vessels, art, and utilitarian objects even after iron came into use (Figure 1.3).

Iron developed in a more haphazard way around the world due to the difficulty involved in mining and shaping iron and, in some cases, a cultural preference for the look and feel afforded by bronze. Ancient peoples first used meteoric iron (an alloy containing nickel), hammering it into the desired shape. This alloy is naturally found in meteors, and did not need to be smelted for use, and some people used it without ever having discovered bronze. However, while iron ore is found throughout the world, iron is rarely found in nature in a pure state. Extracting pure iron from iron ore requires smelting. It is during the smelting process that carbon can be introduced to iron, resulting in steel, a harder and stronger material. Since steel was lighter

1: INTRODUCTION

Figure 1.3 Ancient bronze helmet. Planner/Shutterstock.com.

and cheaper than bronze, and in many cases stronger, it became the preferred metal for weapons, armor, containers, and other objects in many parts of the world. Artisans began to depart from the casting methods that were commonly used for bronze, and began utilizing forging and tempering, developing the blacksmith tradition. More advanced metallurgy and smelting at higher temperatures led to the crafting of new kinds of steel, using different amounts of carbon and nickel and other alloying elements. The swords and weaponry of the middle ages were the result of the materials science developed during this period (Figure 1.4).

Ulfberht, the famous Viking sword featured on the PBS NOVA program [14], is the first known example of fine steelmaking. In fact, it predated similar technologies by nearly a thousand years. While most medieval weapons were made from common wrought iron, the Ulfberht swords—so named because the brand was marked in the blade itself—were made from something called crucible steel. At the time, Vikings would most likely have acquired the technology from trade or conquest activities in Central Asia. Fired at a much higher temperature and made

Figure 1.4 A sample of iron ore. Kletr/Shutterstock.com.

Figure 1.5 Ulfberht swords, forged of crucible steel, were superior in performance to all other swords made during this period. *Photo by Torana, shared on Creative Commons.*

liquid, crucible steel absorbed a much larger amount of carbon, while impurities—such as slag and other particulates—were dissolved. Crucible steel was therefore much stronger and more flexible, and made a much finer (and more deadly) edge. It was able to pierce through most armor (Figure 1.5).

With all of the historical and archaeological emphasis on stone and metal it can be easy to forget about other materials. But metal blades do not cut most efficiently and effectively all by themselves. A handle—usually made of wood or another relatively light, sturdy material—provides grip and leverage, and supports the blade for greater accuracy. The blade must be bound to the handle in such a way that heavy use will not dislodge it. For as long as humans have been using raw materials, we have also been inventing and improving fasteners, fillers, binders, and accessories that, put together, turn a man-made item into a functional thing. While often overlooked, this kind of practical, hands-on experience is an important part of materials science; sometimes it is even more important than the technological advances in the processing of raw materials.

1.3 Other Materials

Wood has been, and continues to be, a basic building block of human architecture and machinery. The Ancient Egyptians built statues and other detailed works of art out of wood. The Chinese timber frame is considered one of the most important contributions to architecture worldwide (Yingzao Fashi). African artisans developed unique carving techniques as they used wood for architectural embellishments. People all over the world made masks, jewelry, and ceremonial items out of wood. Traders harvested and sold unique varieties, like ebony and mahogany, from various parts of the world—many of which are still some of the most valuable materials today. Nearly every seafaring society built ships out of wood, often coating it in pitch or other water resistant sealants to prevent waterlogging. The oldest simple machines—including medieval weapons such as the trebuchet—were made from wood, as were the first wheeled vehicles. The Mississippian culture of North America built enormous pyramid-like religious and residential structures out of river clay and sand, topped with buildings entirely from wood from 600 AD to the 1400s. South Americans used a great deal of wood in their famous architecture as well. It has been, perhaps, the most ubiquitous building material, though it does not preserve well in the archaeological record compared with nonorganic substances. Yet there is no "wood age," perhaps because people have always used wood and will continue to do so. It is a functional, versatile, and beautiful resource, and when used properly, a renewable one as well.

In China, artisans learned how to craft bamboo—an exceedingly light, hollow wood—in ways that are still poorly understood in most of the world. Mass production of paper made from bamboo fibers enabled the development of paper currency and, of course, accessible and cheaply made tablets for writing. The Chinese also invented paper kites, then around 600 AD combined flexible bamboo with silk for the fabric and twine, and sometimes equipped these kites with strings and whistles for musical quality. Kites and other flyers were not just for entertainment, but became useful for military communication, and provided the basis for the science of aerodynamics [15]. Much later, wood was even used to build the first airplane.

Then there are ceramics, fired clays, and glass. Around the same time Americans were crafting wood and clay, artisans in China were developing porcelain, a ceramic material made from processing a specific type of clay. The first artifacts made from these materials include pottery: containers used to carry water and food. Over time ceramics have become more specialized, with new firing techniques and glazing giving rise to a wider variety of materials. The properties of ceramic make it an ideal material for a number of surprising functions. For example, ceramics are used in the auto industry for a wide range of applications including engine parts and brake disks, due to the material's resilience at high temperatures. Ceramics are also sometimes used in the medical field, as synthetic bones or for orthopedics.

Chemistry played an important role in the use and development of new materials. As a study of the composition, properties, structure, and change of matter and energy, chemistry is an ancient field first known as alchemy, which also encompassed metaphysical concerns. Though alchemy had long been practiced around the world, Grecians are credited with coming up with the first basic theory of chemistry [16]. Chinese artisans also developed chemical solutions, devising the explosive potassium nitrate from sulfur, charcoal, and saltpeter (later used in the Chinese cannon). Chemistry was canonized as an empirical academic discipline by Sir Francis Bacon, Robert Boyle, and their colleagues in sixteenth-century Britain.

Another aspect in the use of new materials is fuel sources, which arguably drove the development of metallurgy and, ultimately, the Industrial Revolution. Fuel has always been necessary in the production of ceramics and glazes, metals, and certain stonework. It was the discovery of coal—which replaced wood because it burned hotter and was easier to find—that enabled the Industrial Revolution.

1.4 The Industrial Revolution

Beginning in the late 1700s and lasting into the 1800s, new manufacturing processes, the use of coal as fuel, the development of steam technology, new agricultural methods, and new machines all contributed to the boom in production and innovation known as the Industrial Revolution. Along with machine production came a new concept: the factory. Up until now the makers-of-things were part of the village, or the tribe. They might be a carpenter or a blacksmith or a tailor or even a mad-hatter. They had their workshops, and sometimes there were clusters of artisans—the first cottage industry as it were. But then the concept of the factory developed. The factory enabled not only mass production, but also specialization. Each machine designed a different part: one machine would pour the glass

1: INTRODUCTION

into a mold while another pressed the caps, while yet another applied the caps to the jar. At first, factories still required a lot of human labor too. Mass migration from rural areas to the cities provided that, and people worked alongside machines in factories (Figure 1.6).

Wheeled vehicles—carts, wagons, trains, and later trucks and automobiles—and ships, of course, made relatively fast work of transporting materials from the mine to the factory, and then to the consumer. Aside from enabling a massive increase in metal production—particularly lead, iron and steel—the Industrial Revolution also saw innovations in chemistry, gas utilities and lighting, cement, glass, paper, transportation infrastructure (such as railways), and mining.

The advance of the machine did not result in a betterment or worsening of goods overall. Rather, where once there had been bad artisans and good ones, now there were dysfunctional machines and excellent ones. Depending on the company, the materials used, and quality of the human labor (often related to how well the laborers were treated), factories produced a wide range of products from shoddy to quality-made. Overall, consumers benefitted because there was better access to a wide variety of products including clothing, household goods, transportation, and fun things like toys and entertainment, and advances in packaging and distribution allowed for low cost access to quality food and medicine.

Most things being made by machines had another effect: the resulting objects looked and behaved the same. It became popular for people to buy what was fashionable. Consumers wanted to emulate celebrities—and each other. Having a brand name became a mark of social status. Originality

Figure 1.6 Antique illustration of Aubin forging mills, France. *Original, from drawing of Forest, published on L'Illustration, Journal Universel, Paris, 1860.*

now took extra effort and had to be customized, and so it became the mark of the ultrawealthy and the eccentric. This changed the way things fit into people's lives. People had a new relationship with objects, and by extension, a new relationship with the artisan or manufacturer. The Age of Mass Production had arrived.

1.5 Mass Production

As factories became more and more specialized, the machines used in the factory began to take center stage. Humans designed the machines, turned them on, maintained and cleaned them. The saying "Never buy a car that was made on a Monday"—insinuating that the workers might have had a little too much fun over the weekend—became less relevant. Now machines could do things that could not be done by hand. They could make and move bigger and heavier parts, like ship propellers. All made the same way, products became more reliable. They were all the same size and could fit in the same spaces. For example, doors made out of the same materials and all the same way would fit into the same size doorways, resulting in standardization of building practices and, therefore, cheaper prices all around. More complicated devices could be put together in an assembly line, allowing for replaceable parts to be used.

It was Henry Ford who sponsored and championed the assembly line, promoting an industry standard known as "Fordism," which prioritized high wages for employees and cheap, accessible products for consumers. Ford also implemented the first vertically integrated factory, with steel and glass developed in-house. Thus Ford developed the first affordable automobile, the Model T, which was introduced in 1908. As part of the standardization process, Ford later had the Model T painted only in black because that was the cheapest color to use. He famously said: "Any customer can have a car painted any color that he wants so long as it is black" (Figure 1.7).

In *The Work of Art in the Age of Mechanical Reproduction*, the philosopher Walter Benjamin discussed what happens to the originality of a work, and its social impact, once it can be easily reproduced. If we take liberties in paraphrasing Benjamin, we can say that the author of a work lost face once that work could be reproduced en masse. Extending the analysis beyond works of art, we can say that the machine entered the realm of production as a primary interlocutor, and the value of all things based on their reproducibility became one of the driving forces behind their initial design.

1: INTRODUCTION

Figure 1.7 Ford Model T factory.

But as it turns out, the role of the artisan or designer—the person who makes a thing—is still every bit as important in the age of mass production. One of the reasons for this has to with materials science.

1.6 Materials Science

One of the most important things about designing a product is choosing the right material for its manufacture. In the human history of making things, we have managed to come up with hundreds—if not thousands—of different raw materials to choose from, as well as a number of varieties of each material having to do with where it came from, the specific physical and chemical properties of that particular source, and its method of manufacture or processing.

In the old days, if an artisan made a clay bowl that broke, the user would be unhappy. If it happened again, the user would not buy any more bowls from that artisan. The same with a blacksmith: one did not become a respected blacksmith by making shattering swords. But when a manufacturer produces a part that breaks, the consumer (and often the company too) shrugs it off as a manufacturing defect, or simply as a bad design. Rarely do they consider the very real possibility that they are simply using the wrong material.

Choosing proper materials has always been an important part of the creative process, one that requires learning from those who are not only knowledgeable, but who are also experimenting, and building on existing theory and practice. In the era of mass production, proper material selection takes on a whole new level of importance because of the degree of specialization required. However, it was not until 1955 that the study of the properties of materials became formalized when faculty at Northwestern University in Illinois founded the first Materials Science Department [17]. As materials science has developed and become formalized, professional societies have formed to help guide practitioners in their respective fields. These include the following:

The American Society for Materials

The Aluminum Society

The Association for Iron and Steel Technology

The Pulp and Paper Association

The Concrete Institute

Professionalization has become necessary because the process of material selection has changed from a task based on conversation and familiarity, to one of mathematical analysis. And often, one needs an advanced degree just to understand the math.

Plastics entered the scene during and after the professionalization of materials science, so unlike older materials—like stone, steel, and the ever ubiquitous wood—it lacks that history of apprenticeship. There is no oral tradition where knowledge and insights and sage wisdom about plastics is passed from generation to generation. However, a person who works with plastics needs to have an appreciation for their history, their variety, and the attributes that make them so functional, versatile, and yes—even beautiful.

1.7 The Plastics Age

Plastics are organic materials, that is, they are composed of chemical compounds containing carbon. These compounds are linked together to form molecules known as polymers. While we may think of polymers as being new and high tech, the reality is that they have been around forever. They exist in nature in many forms. Latex rubber, made from the sap of the rubber tree, is the most obvious example. Rubber tapping is an age-old tradition practiced around the world and usually passed down through generations. Even today, families and communities still keep groves of rubber trees, carefully tapping them for the viscous sap, plugging up the holes, and working their way up the tree as each spot heals over.

: Introduction

Rubber is not the only naturally occurring polymer; all plants contain cellulose in their cell walls. Cellulose is a polymer based on carbohydrate compounds. Egg and blood proteins—long used to make paint, glue, and textiles—are also naturally occurring polymers. Naturally occurring polymers have always been used as fillers, additives, adhesives, and colorants. Some of these polymers are still in use today, and are resurfacing in today's plastics industry as a viable material for use in bioplastics, and the sustainable rubber trade is still going strong.

Today, much of what we think of as plastics are synthetic materials, which came about as a result of chemical synthesis. The first synthetic, laboratory-born polymer came about as a replacement for ivory. In the 1800s elephants and other ivory-bearing animals were dying out because of the prolific use of ivory, for everything from the boning in corsets, to combs, to piano keys, to billiard balls. Inventor John Wesley Hyatt made celluloid, meaning *like cellulose*, by experimenting with cotton. Similar to naturally occurring polymers, it was more versatile, although it had the drawback of being extremely volatile and explosive. It did not quite make good billiard balls—not yet—because celluloid balls were too loud when they hit each other, and factory production was extremely hazardous. But celluloid replaced ivory for other products, most notably film. The success of celluloid led to development of other materials, such as Bakelite (whose rather daunting chemical name is polyoxybenzylmethylenglycolanhydride) and other phenolics. Bakelite was the first thermoset plastic ever made, and was invented to replace shellac, a polymer derived from insect excretions. This phenol–formaldehyde solution had the advantage of being more versatile, and much less volatile, than celluloid [18,19].

Further advances in polymer development were led by the DuPont company. Nylon (polyamide), invented in 1935, was fabricated as a silk replacement, and was introduced at the World's Fair as a textile in 1939. The material, a thermoplastic, was lightweight and cheap to produce. It was also resistant to heat, mold, mildew, and abrasion. Since it was also very strong, it replaced metal in the manufacture of many mechanical parts and tools. One of the most versatile materials ever made, it was used to make everything including parachutes, stockings, tents, musical strings, sausage casings, and tires. Quite literally, nylon changed the world. It was nylon production that led to the development of more specialized polymers, like Saran Wrap (polyvinylidene chloride), whose impermeability makes it ideal for food storage. By the 1950s, synthetic polymers had become a staple of everyday life (Figure 1.8).

Today, the body of knowledge about plastic materials and their use is vast. There are hundreds of thousands of people who work in the plastics industry.

Figure 1.8 Nylon parachute.

In the United States, plastics represents the third largest manufacturing sector. There are hundreds of universities around the world offering degrees in Polymer Chemistry and Plastics Engineering. There are dozens of professional societies dedicated to various areas of plastics technology. There are thousands of technical books, and hundreds of thousands of technical papers on plastics processing, plastics design, and plastics engineering. Yet, on a societal level, even a personal level, most people are ignorant about plastic materials—and even think about them negatively because they are synthetic. But plastic is not the first ubiquitous, and indispensable, synthetic material.

1.8 Plastics—The Other Synthetic Material

Many people think of plastics and other synthetic materials in a negative light. In common language, plastic has sometimes become a derogatory term people use when they want to imply that something is cheap, superficial, or artificial. This attitude is linked with fear of chemicals, or "chemophobia" [20–22].

To better understand this mentality, let us take a look at the etymology of the word "synthetic." Originally, it comes from the Greek *synthetikos*, which means "skilled at putting together, constructive." It is generally understood to be something that is not found in nature, meaning that one has to put together elements of other things to make it. Steel is often idealized as "the real thing," but it is man-made, too. It is a metal alloy, composed of manipulated iron and other elements. The reality is that steel is a synthetic, man-made material.

Plastic is also a synthetic, man-made material, and is often considered the antithesis of natural. But is natural always better? There are lots of natural materials that are toxic—lead, arsenic, mercury, asbestos, hemlock.

Both plastics and steel are man-made, but that does not make them unnatural. They are made from elements found in nature, by people who as carbon-based life forms are not only part of nature, but have been working with natural elements and modifying those elements since early hominids walked the earth. Just as a metals expert constructs and alters metals, so does a polymer chemist synthesize compounds from organic elements. In reality, chemicals and synthetics are a result of a creative process—or, in other words, people making stuff.

So whatever you think about plastic, it is definitely not artificial or fake. Artificial implies that something acts as a substitute for "the real thing"—like artificial flavors. While plastic is sometimes a substitute for another material, plastics are unique materials with unique properties that make them ideal for a wide variety of uses.

Now, let us look more specifically at the word plastic itself. Before the word *plastic* (from the Greek *plastikos*) was used to name the material we know today, it was used to describe anything that could be easily molded or shaped. We still use it as an adjective: clay has plastic qualities, for example. The substance we call plastic was so named for its very physical characteristics—plastic materials can be easily molded or shaped.

While there are concerns around the environmental implications of plastics use, some legitimate and some not, a quick look at the history of the material reveals another side to the matter. Plastics were originally invented to replace ivory, that extremely "natural" substance whose use today is responsible for several present-day extinctions of large mammals, such as recently, the Black Rhinoceros. Like most materials, plastics can be more responsibly and aptly used when better understood. The reality is that plastics are all around us, and will continue to be important to our daily lives. Proper understanding of these materials will lead to innovative and sustainable use and development.

1.9 Plastics Material Selection

As discussed earlier, proper material selection is a critical component of any manufactured product. This is especially true with plastics. Traditionally, engineers and manufacturers have used a combination of methods for selecting plastics materials, including a comparison of published property data, recommendations from material suppliers, and good old trial and error. The process can be difficult, time-consuming, expensive, and often frustrating.

What is often lacking in this process is a fundamental understanding of plastics as a raw material—a substance used to make things; one that requires creativity and consideration. For example, the blacksmith had to acquire a lot of detailed knowledge about materials to make a good sword, which is really a very advanced product. Many people who work with plastics think it is enough to use "SKP"—just *some kind of plastic*. This results in a bad quality product, just like a blacksmith making a brittle sword containing too many impurities. How can you make something out of plastic if you do not understand the material? This book is designed to fill that gap, by describing the characteristics of these unique materials and providing a guide for effective plastics material selection.

1.10 How This Book Can Help You

This book is for a diverse audience: engineer, manufacturer, industrial designer, supply chain professional, architect, scientist, student, physician, teacher, student, politician, policymaker, artist. Students and educators can use it both as a resource for information and a guide in the classroom. Psychologists, historians, and anthropologists, who look at the way in which each individual, and the species, grows through creative interactions with objects and materials in everyday life, will find interesting lessons in the story of plastic. Astrophysicists, physicists, and chemists will gain insight toward the application of the material by knowing more about its origins, properties, and potential. Medical professionals may want to make products for patients that are more marketable, usable—and even fun. Policymakers—people who legislate, regulate, and make policies and procedures—may become better informed to make good decisions.

Last, but far from least, creative technical specialists—people who are in the business of making things—can use this book to guide their work. Designers, engineers and other specialists need to know how to choose the right material for their products. While the right material may or may not be plastic, learning about the advantages of plastics and how to choose them will open your horizons and give you more flexibility in your field. This is, of course, also useful information for supply chain professionals: those in the trenches of the manufacturing world, getting the job done everyday and forming the backbone of the industry.

While an ambitious project, it is a necessary one. Even though we are in the midst of The Plastic Age, the plastics industry is transforming, undergoing some critical self-evaluation, and feeling more than a few growing pains. Plastics are not going away; if anything they are becoming more prevalent. The industry is growing and becoming more innovative. 3D printing has introduced a new level of efficiency to the market. New material technologies, such as bioplastics, are garnering more attention due to a depletion of natural resources and a sense of urgency about the effects of global warming. Professionals in all industries need to become familiar enough with plastic materials to understand their traditional use and manufacture, their history as a material, their untapped potential, and their role in the future.

References

[1] S.P. McPherron, Z. Alemseged, C.W. Marean, J.G. Wynn, D. Reed, D. Geraads, R. Bobe, H.A. Béarat, "Evidence for stone-tool-assisted consumption of animal tissues before 3.39 million years ago at Dikika, Ethiopia." Nature 466 (2010) 857–860.

[2] Annalee Newitz, What Modern Humans Can Learn from the Neanderthals' Extinction, Popular Science, May 16, 2013.

[3] Ker Than, Neanderthals, Humans Interbred – First Solid DNA Evidence, National Geographic Daily News, May 6, 2010.

[4] Carl Zimmer, Interbreeding with Neanderthals, Discover Magazine (March 2013). http://discovermagazine.com/2013/march.

[5] Lyn Wadley, Two 'moments in time' during middle stone age occupations of Sibudu, South Africa, Southern African Humanities 24 (2012) 79–97.

[6] University of the Witwatersrand, Modern Culture 44,000 Years Ago: Human Behavior, as We Know It, Emerged Earlier than Previously Thought, ScienceDaily, 2014 (accessed May 25) www.sciencedaily.com/releases/2012/07/120730155049.htm.

[7] Roberto Maggi, Mark Pearce, Mid fourth-millennium copper mining in Liguria, north-west Italy: the earliest known copper mines in Western Europe, Antiquity 79 (303) (2005) 66–77.

[8] Petar D. Glumac, Judith A. Todd, Early metallurgy in southeast Europe: the evidence for production, in: Petar D. Glumac (Ed.), MASCA: Recent Trends in Archaeometallurgical Research 8 (1) (1991) 5–19.

[9] Michael Given, Mining landscapes and colonial rule in early-twentieth century Cyprus, Historical Archaeology 39 (3) (2005) 49–60.

[10] Eugenia W. Herbert, Red Gold of Africa, University of Wisconsin Press, Madison, 1984.
[11] Guy Gibbon, Old copper in Minnesota: a review, Plains Anthropologist 43 (163) (1998) 27–50.
[12] Claire G. Goodman, Copper Artifacts in Late Eastern Woodlands Prehistory, Center for American Archaeology, Northwestern University, Evanston, 1984.
[13] Dragana Antonovic, Copper Processing in Vinča: New Contributions to the Thesis about Metallurgical Character of Vinča Culture, Archaeological Institute, University of Belgrade, 2002.
[14] Peter Yost, Secrets of the Viking Sword, PBS NOVA, 2012.
[15] John David Anderson, A History of Aerodynamics and Its Impact on Flying Machines, Cambridge University Press, 1997.
[16] Lois Fruen, The Real World of Chemistry, sixth ed., Lois Fruen Kendall/Hunt Publishing, 2002.
[17] Megan Fellman, World's First Materials Science and Engineering Department Turns 50, Northwestern University News, October 18, 2005.
[18] Susan Frankel, A Brief History of Plastic's Conquest of the World, Scientific American, May 29, 2011.
[19] Stephen L. Sass, The Substance of Civilization: Materials and Human History from the Stone Age to the Age of Silicon, Arcade Publishing, 2011.
[20] M. Lorch, Why Do People Hate the Word 'Chemicals'? BBC News Viewpoint, 2013. http://www.bbc.com/news/magazine-25103941.
[21] K. Sanderson, What are you afraid of? Chemistry World (October 29, 2013). http://www.rsc.org/chemistryworld/2013/10/chemophobia.
[22] Chemophobia Podcast. http://www.bbc.co.uk/programmes/b03jdw70.

2 Why Use Plastic?

2.1 Introduction

The word *plastic* has its origin in the Greek word *plastikos*, meaning to shape or to form. Today, plastic has many different meanings. In common use, it is an all encompassing term, often meaning anything and everything (and sometimes even nothing). However, in the technical world, plastic has some very specific meanings.

In engineering, plastic is used to describe a specific type of mechanical behavior in a material. This behavior—commonly called plastic deformation—occurs when a material deforms and undergoes a permanent change of shape in response to external forces. It is different from elastic behavior, which occurs when a material changes shape in response to an external force, but then returns to its original shape when the force is removed. As an example, think of a paper clip. Under normal use, the clip bends slightly when it is slipped over a few sheets of paper. When the clip is removed, it returns to its original shape. This is elastic behavior. However, if you grab the ends of the paper clip, and twist them, they will bend and deform, and when you let go of the ends, the clip will have a different shape. This is plastic behavior, and the nonreversible change in shape is called plastic deformation. Most materials—including wood, stone, steel, concrete, even bubble gum—will exhibit a certain amount of elastic behavior, and a certain amount of plastic behavior, depending on the circumstances and the forces involved (Figure 2.1).

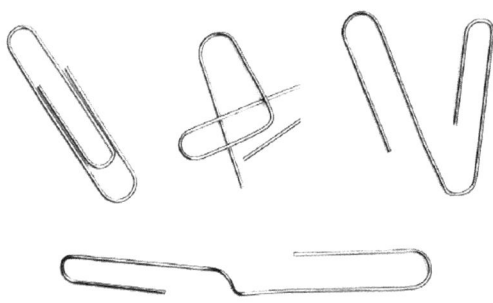

Figure 2.1 Paper clips, as manufactured, and bent and distorted. Olga-Kovalenko/Shutterstock.com.

In manufacturing, the term is used to describe materials that can be molded—either by pouring a liquid into a mold and letting it cure into a solid, or by using heat and pressure to transform a malleable material into another shape. Collectively, these materials are referred to as plastics, or as plastic materials. Fortunately—and perhaps not coincidentally—most plastic materials also exhibit plastic behavior in use. While there are plastic materials that can be found in nature (such as latex rubber), most plastic materials are man-made, synthesized by combining various organic source materials.

Regardless of the technical distinctions, in common use, the word *plastic* is most often used to describe plastic materials. So when we ask the question, *Why Use Plastic?* what we are really asking is, *What makes these materials so special?* And while that may seem like a very simple question, the answer or perhaps I should say the answers are quite complex. To effectively answer the question, we need to know more about plastic materials, their chemistry and behavior, the ways that they can be processed, their niche in the world of design and manufacturing, and most importantly, what makes them unique.

2.2 Plastics as Raw Materials

All plastic materials are polymers, or in other words, they are made of molecules consisting of many parts. The word comes from the Greek word *polumeres* (*polus*, meaning many, + *meros*, meaning part or component). Plastic molecules typically consist of atoms and small compounds of carbon, hydrogen, nitrogen, etc. The atoms and compounds are first bonded together into a small molecule, called a monomer. These monomers then bond to one another—usually in a repeating, linear fashion—to form a chain. This chain is a polymer molecule. The length of the chain depends on a number of variables, including the chemistry of polymer itself (Figure 2.2).

Some polymers are naturally occurring substances, while others are synthetic (man-made). Regardless of their origin, polymer molecules are

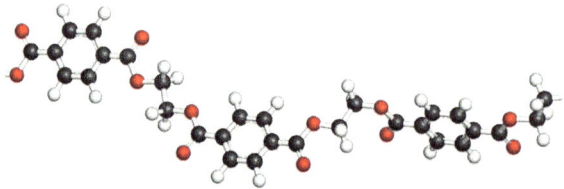

Figure 2.2 Polymer molecule. molekuul.be/Shutterstock.com.

normally quite stable. The bonds within the chain are strong and durable, and when you collect a large quantity of polymer chains together, they have an affinity for one another via interchain attractions. Think of the collected chains as a polymer matrix. The matrix may be a solid or liquid at room temperature, depending on the chemistry. However, in order to be useful, we normally want a polymer that is solid at room temperature (and often at elevated temperatures as well).

Plastic materials are broadly sorted into two different categories: thermosets and thermoplastics. Their differences are described below. Interestingly enough, for most of us within the plastics industry, when we talk about plastic materials, the terms *plastic* and *thermoplastic* are used interchangeably. However, when we speak of thermoset materials, we use the word *thermoset*. Regardless of the terminology, it is important to remember that these are distinctly different categories of materials. While the in-use behavior of a thermoplastic material may be identical to the in-use behavior of a thermoset material, they will have significant differences in their chemistry, as well as in their processing and handling.

2.2.1 Thermosets

Thermosets (perhaps more correctly described as "thermosetting plastic materials") are materials that are liquid or malleable in their initial state but are then converted into a solid form. The conversion process involves a chemical reaction typically triggered by heat, oxygen, UV light, a reagent material, or a catalyst. Regardless of their initial state, the important thing to remember about thermosets is that the conversion process is irreversible. It is kind of like baking a cake: you mix the batter, you pour the batter into a baking pan, and you bake the cake. Once the cake is baked, you cannot turn it back into batter.

Thermosets are basic building blocks of a wide variety of products that you come across in daily life, including paints, lacquers, varnishes, adhesives (such as epoxy), caulks, and all kinds of rubbery stuff. Most of this rubbery stuff is in the form of vulcanized rubber, which is used in shoe soles, tires, and hockey pucks.

2.2.2 Thermoplastics

Thermoplastics are plastic materials which melt upon heating, and solidify upon cooling. During the molten phase (which usually involves both heat and pressure), the materials are malleable and can be easily formed into another shape. Upon cooling they become a solid, and retain that shape. The melting/solidifying process is fully reversible, and most thermoplastic materials can be molded again and again and again.

Thermoplastics are ubiquitous in modern life. They are used in products that one normally thinks of as being made out of plastic—plastic soda bottles, CD discs, children's toys, cell phone housings, grocery bags, prescription bottles, rubber duckies. However, thermoplastics are also used in products that you might not realize are made out of plastic—automobiles, washing machines, lawn mowers, chain saws.

Added together, thermosets and thermoplastics represent a unique kind of raw material: plastics. And while they do have their differences, on many levels they are the same: they are made of polymer molecules, they exhibit plastic behavior in use, and they can be molded.

2.2.3 Key Differences

At first glance, the difference between thermosetting plastic materials and thermoplastic materials seems pretty simple: one you can heat and melt and reprocess, the other you cannot. But there are some additional differences between them, some of which are subtle, and some not. To understand those differences, we need to look at these materials on a molecular level.

As described earlier, plastic materials are made out of polymer molecules, which are dispersed within an overall polymer matrix. In a thermoplastic material, when we apply heat the polymer matrix becomes soft and malleable. Depending on the material—and how much heat and pressure is applied—it may become a true liquid, or it may turn into nothing more than a blob of hot goo. In either case, the material can then be formed into a new shape using a mold or tool, and then, once the heat is removed, the material returns to a solid form. The polymer chains are once again dispersed within the matrix, and the material regains its strength and rigidity.

In a thermosetting material, something very different happens. During the process of being formed, a chemical reaction occurs *between* the chains of the individual polymer molecules. As a result, chemical bonds are then created between the atoms and compounds on one polymer chain and the atoms and compounds on another polymer chain. These types of bonds are described as cross-links. The cross-link bonds material may be weak, or very strong—sometimes even stronger than the bonds within the polymer chain itself.

The chemical reaction itself can be triggered by a variety of means. Sometimes it happens when two reactive polymers are mixed together, as with epoxies, and sometimes when the polymer is exposed to oxygen, as with silicone caulk, or UV light, or a catalyst. Most of the time, this chemical reaction occurs when the polymer is exposed to heat, and more often than not, the chemical bonds that form between the polymer chains, the cross-links, are permanent. They are cured, or set. Hence the term, *thermoset*.

The important thing to remember is that the polymer chains are once again dispersed within the matrix. All of the original *intramolecular* bonds are still in place (those within a given polymer chain) but there are now additional *intermolecular* bonds between different polymer chains. The strength of these intermolecular bonds, and their quantity, will affect the properties of the material in its final state.

In their precured state, thermoset materials can take a variety of forms. They might be a solid, a liquid, a gel, a paste, or anything in between. They are typically packaged so that they will retain that form until the activation process, when the cross-linking process will be initiated. And while the cross-linking process may sound like something out a science fiction novel, it is actually a process most of us experience quite often. We squeeze a dab of cyanoacrylate out of a tube, and it then reacts with oxygen, and hardens into superglue. We squeeze some silicone out of a caulking gun, and use it to cover the seams in our bath tub. After several hour of exposure to oxygen, it solidifies into a flexible sealing material. Or we mix two dabs of molasses-like goo and make an epoxy adhesive. Or, we mix two different vials of a corn syrup-like material, and then pour them into a mold—perhaps over a mat of woven glass fibers—and then let it harden into a surfboard, a rain gutter, perhaps even a piece of jewelry (Figure 2.3).

This cross-linking process is also utilized on an industrial scale—and not just for glues, caulks, and sealing materials. In the aerospace industry, thermoset materials are often used in the fabrication of composite structures, where fibers made of pure carbon—either in long strands or woven into fabric—are saturated with a liquid resin which hardens (cures) into a solid matrix, trapping and bonding with the fibers

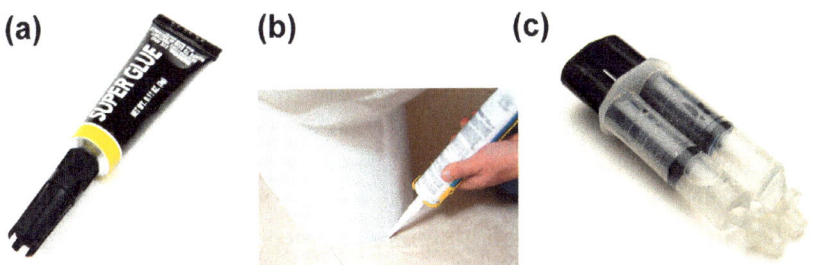

Figure 2.3 Some common examples of thermoset materials, (a) superglue, (b) silicone caulk, (c) 2-part epoxy adhesive. DavidBrimm/Shutterstock.com, lightwavemedia/Shutterstock.com, *and* CraigWactor/Shutterstock.com.

to make a composite material that offer the highest strength-to-weight and stiffness-to-weight of any material known to man. Working with thermosets is similar to working with many other structural materials, such as cement, which is mixed with water and then hardens into concrete. You get all your raw materials together, you set up your system for fabrication, and then you go to work—with one caveat. You get one chance to get everything right. Otherwise you throw everything away and start over.

As a general rule, thermoset materials kick ass over thermoplastic materials. Thanks to the intermolecular bonds, thermoset materials are usually stronger, stiffer, and more resistant to chemical attack than thermoplastic materials. And, since the intermolecular bonds are more or less permanent, the matrix itself would not soften into a blob of hot goo when heat is applied. However, thermoplastic materials do have some unique characteristics, including some significant advantages over thermoset materials, especially when it comes to the world of plastics processing.

2.3 Plastic Processing Technologies

The processing of plastic materials involves several distinct stages: the creation of the raw materials; the storage, handling, and transportation of these raw materials; the transformation of these raw materials into finished parts (or intermediate products); and the storage, handling, and transportation of the finished parts (or the intermediate products).

This section is focused on the transformation process, where plastic materials are converted into finished products. The intent of this section is NOT to be a comprehensive textbook on injection molding, or any other plastics processing technology. Rather, it is intended to provide enough guidance to allow the reader to evaluate whether a given process is suitable for producing parts made of plastic for a given end-use application.

One of the interesting things about plastics is that they can be processed into finished parts using a wide variety of technologies. As with all manufacturing processes, each plastic processing technology has advantages and disadvantages, along with limitations and constraints, including what materials can be used with that process. Quite often, the best material for a given application is one that is selected specifically to take advantage of a given manufacturing process.

2.3.1 Plastics Tooling

Most plastic processing technologies involve the use of tooling. Tooling is a generic term, used to describe the tools and machinery used in a given manufacturing process. It is NOT the manufacturing equipment itself, but the specialized aspects of that machinery that are used for a specific project. As an example, an automobile assembly plant may have machinery for robotic welding. The robots would be considered equipment, but a custom-made fixture that is used to support the frame of specific model of a specific car would be considered tooling. Tooling is typically customized for a specific operation in a specific project, and is rarely useful when that project is complete.

The most common tools in plastics processing are molds. A mold is a premade object, which is used to guide the final form or shape of another object. Molds can be open, like a baking pan, allowing for soft or molten material to be poured or draped over the mold surface. After the material cools or solidifies into its final shape, it is then removed from the mold, and the mold is used again. Molds can also be closed, with internal hollow spaces, allowing soft or molten material to fill these spaces. Closed molds are usually made of several parts, and fabricated in such a way that the mold can be opened, allowing for part removal after the plastic material has solidified.

In plastics processing, molds are typically made out of steel, but can also be made out of cast iron, aluminum, ceramic, other plastics, even wood. The material(s) used in a given mold depends upon its design, the processing technology being used, the plastics being processed, the size, shape, and quantity of the parts being fabricated, the desired precision for those parts, and how long the mold is intended to be used (Figures 2.4 and 2.5).

Figure 2.4 A baking pan for making muffines—an example of an open mold. ffolas/Shutterstock.com.

Figure 2.5 A common injection mold—an example of a closed mold. HaslamPhotography/Shutterstock.com.

2.3.2 Plastics Forming

Plastics forming is a small but important subset of plastics processing. In plastics forming, plastic materials are simply formed into a new shape. Unlike more traditional molding methods, where the material must be in a liquid or semiliquid state, forming only requires that the material be soft and ductile enough to be formed into a new shape. Forming is an age-old technique, and has been used with all kinds of materials, in all kinds of ways: wet clay being slapped into shape, wood being steamed and bent, plant tissues pounded into a paste, and then flattened and dried to make paper. The forming of plastics is not that much different, except they may need a different kind of processing, along with some help (heat, pressure, etc.).

Plastics forming usually involves the use of open (or one-sided) molds, or jigs or fixtures or dies, or a combination thereof. Some common methods include vacuum forming and pressure forming, another is plastic welding. Extrusion of plastics is a similar processing technology, involving elements of both forming and molding.

2.3.3 Vacuum Forming

Vacuum forming is a variation of a classic forming technique, where a sheet of material is heated slightly (not melted) to make it pliable, and then placed over a mold. In the classic technique, the material simply drapes over the mold, as a result of gravity and the mechanics of the material. (Think of a piece of leather, and a wooden shoe mold.) In vacuum

2: Why Use Plastic?

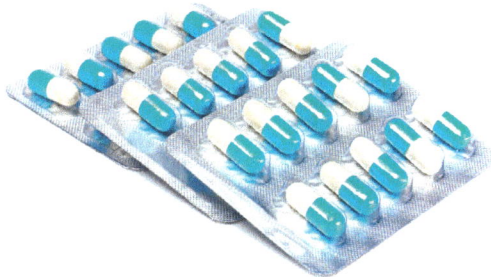

Figure 2.6 A common blister pack, made via vacuum forming. zcw/Shutterstock.com.

forming, not only does the softened plastic drape over the mold surface, but a vacuum is also applied to the top surface of the mold, and as a result, the softened material is sucked down onto the mold and closely follows the mold's shape. It is kind of like blowing a gigantic bubble out of bubble gum, and then, after the bubble breaks, you inhale deeply, and the film of bubble gum forms a contour over your face (Figure 2.6).

When the plastic is cooled, it is removed from the mold, but retains the shape of the mold. It is then cut from the sheet, sometimes with some additional steps to make the final part. The molds can be made from a number of materials, but since the pressures are low, the structural requirements are not that demanding. It is a process well suited for the production of large, reasonably flat parts that do not have a lot of detail, such as a plastic lid for a large refuse container.

Materials:	*Some thermoplastics*
Equipment cost:	*Low–moderate*
Tooling cost:	*Low*
Part cost:	*Low–medium*

2.3.4 Pressure Forming

Pressure forming is similar to vacuum forming, but instead of using a vacuum to suck a heated plastic sheet over a mold, pressure is used to push or force the plastic sheet down over the mold. Since higher pressures are easier to achieve than high vacuums, more force can often be applied to a sheet—and it can follow the mold form more closely. As a result, pressure forming often provides greater detail in the finished product. Sometimes pressure can also be used in conjunction with a vacuum (Figure 2.7).

Figure 2.7 Broken bubble gum bubble. ToddTaulman/Shutterstock.com.

Pressure forming is also like having your face covered with a broken bubble of bubble gum, except now your cousin is holding you down and making sure the bubble gum follows every nook and cranny of your face, nose, cheeks, and mouth. Pressure forming equipment and molds may be a little more expensive than vacuum forming, but in general the costs are similar.

Materials:	Some thermoplastics
Equipment cost:	Low–moderate
Tooling cost:	Low
Part cost:	Low–medium

2.3.5 Plastic Welding

Plastic welding is a unique subset of plastics forming. It is typically used to assemble parts—not to repair them. Unlike the traditional process used in welding metal, where a welding rod is used to add metal to the base structure, in plastic welding, only the original parts are used—no additional plastic is added.

The welding process itself requires heat. This heat can be applied directly (with a heat gun, soldering iron, or hot plate) and can also be generated through friction. Two cylindrical parts can be rotated against each other (spin welding), flat parts can be rubbed against each other (vibrational welding), and some parts can be made to vibrate on a microscopic level through high frequency acoustics (ultrasonic welding).

Welding can only be done with thermoplastic materials. And while most thermoplastics can be welded, the strength of the bond varies considerably from material to material. The typical application welds together two parts made from the same material. Sometimes parts made from different materials can be welded together, provided they are chemically compatible.

Materials:	*Most thermoplastics*
Equipment cost:	*Low–moderate*
Tooling cost:	*Low–moderate*
Part cost:	*Low–moderate*

2.3.6 Extrusion

Extrusion is a general term for a forming technique, where a material is pushed (or pulled) through a die to create a solid material with a specific cross-sectional shape. Extrusion is also used to describe the transfer of semisolid materials from a pressurized area, out through an orifice (a small opening), into the air or onto another substrate, in a smooth continuous manner. (Think of putting a strip of toothpaste on your toothbrush, or a line of caulk on a bathroom tile.) Extrusion is normally a continuous process, although it can also be done in a batch mode (Figure 2.8).

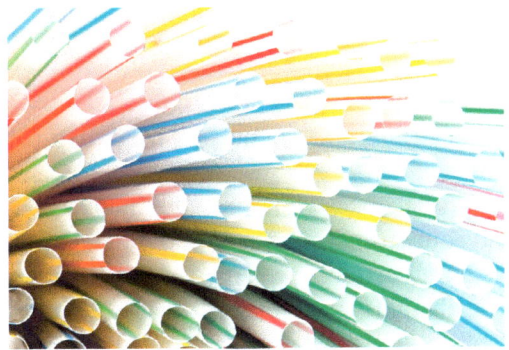

Figure 2.8 An assortment of extruded drinking straws. Drozdowski/ Shutterstock.

Both methods of extrusion are widely used in the plastics industry, although mostly with thermoplastics. Extrusion is how the pellets for most thermoplastic resins are created. The molten resin is forced though a die, creating a continuous rod-like filament. As the extruded filament cools, the pellets are cut from the solid rod.

Materials:	*Most thermoplastics*
Equipment cost:	*Low–moderate*
Tooling cost:	*Low–moderate*
Part cost:	*Low–ultralow*

2.3.7 Plastics Molding

Molding is the primary method of converting plastic materials into finished products. Another age-old technique, humans have molded all kinds of materials, in all kinds of ways, including bricks, glassware, and all sorts of metal parts. Plastics molding usually involves the use of closed molds, and usually involves a combination of heat and pressure.

2.3.8 Casting

Plastic casting is a processing technique, where the material is first converted into a liquid form, and then poured into a mold. The liquid resin fills the mold, and then hardens.

Casting is primarily done with thermoset materials, where the initial form of the material is a pure liquid (they are not heated and melted). In this scenario, the resin is at room temperature, but it must be mixed with a curing agent. It is poured into the mold under normal pressure. The molds are commonly made of latex rubber, silicone rubber, or other similar low-cost materials, and can only be used for a limited number of castings.

Casting can also be done with some thermoplastic materials, but it requires heat and some use of pressure. The pressure may be applied by centrifugal force, or an external pressure source. Resin casting is used to make all kinds of fiberglass products, and is often used to produce dolls, toy models, and other collectibles. Unit cost per finished product is usually higher than other molding technologies. Production quantities are usually low (at least per mold).

2: Why Use Plastic?

Materials:	*Thermosets, some thermoplastics*
Equipment cost:	*Low*
Tooling cost:	*Low*
Part cost:	*Medium*

2.3.9 Compression Molding

Compression molding involves taking a slug of material, and then placing it in an open mold. The mold closes under pressure, and the slug is squeezed and fills the mold cavity. Heat is applied to the mold, and the material goes through a curing process. The mold is opened, and the cured part is removed. A simple analogy would be the making of waffles in a waffle iron (although the waffle batter is more liquid, and there is no pressure applied) (Figure 2.9).

Figure 2.9 An antique waffle iron, the culinary equivalent of compression molding. SteveHeap/Shutterstock.com.

Compression molding is suitable only for thermoset materials. The slug itself must be thoroughly mixed before it is inserted in the mold. The molds are usually made of steel or cast iron, although they could be made of aluminum or other metals. The pressures involved are usually higher than what is used for casting, but much lower than with the various injection molding processes. Parts can be small or large. Compression molding is often used to produce swim fins, rubber boots, and auto parts.

Materials:	*Thermosets*
Equipment cost:	*Moderate*
Tooling cost:	*Low–moderate*
Part cost:	*Moderate*

2.3.10 Rotational Molding

Rotational molding is similar to compression molding, in that you start with an open mold, but instead of placing a slug of material in place, you use raw resin pellets. The mold is then closed and secured. The mold itself is attached to a shaft, which not only rotates about its axis, but the entire shaft is rotated around another axis. The net effect is that the mold is rotated and spun around in the air. As the pellets inside tumble around, the mold is moved into a gigantic oven, and as the mold slowly heats, the pellets melt and the sticky resin coats the entire inner surfaces of the mold. After a predetermined time, the mold is removed from the oven and allowed to cool. The mold is then opened, and the part is removed (Figure 2.10).

Figure 2.10 Rotationally molded parts usually have a thick skin and a hollow core, just like a chocolate Easter bunny. Jules_Kitano/Shutterstock.com.

A rotational molding machine looks somewhat like a crazy ride at an amusement park. The molds are usually made of iron or steel, but since there is no pressure involved, they can be simple shells. The process is suited only for certain materials, those that will melt and coat the mold, and also withstand the heat for the full duration of the molding cycle (which can be several minutes). It is well suited for making simple, hollow parts (things like chocolate Easter bunnies and rubber duckies).

Materials:	*Some thermoplastics*
Equipment cost:	*Moderate*
Tooling cost:	*Moderate*
Part cost:	*Moderate–high*

2.3.11 Injection Molding

Injection molding is one of the primary means of fabricating plastic parts. It is a complex process that consists of several phases. It begins by taking plastic pellets, which are then metered under carefully controlled conditions into a section of the molding machine known as the barrel. The barrel is basically a long tube, and in the center of the tube there is a long fluted rod known as the screw. The screw is then slowly rotated, and the flutes of the screw squeeze and shear the plastic pellets under very high pressure, pushing them forward, melting them in the process (Figure 2.11).

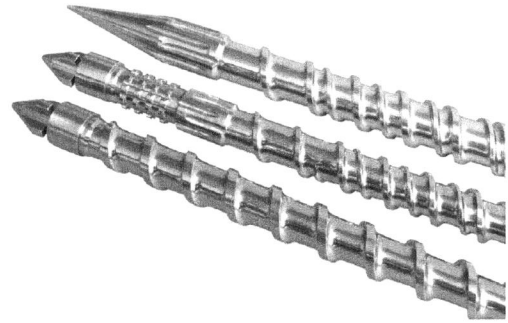

Figure 2.11 Some different screws for injection molding. Valueho/Shutterstock.com.

At the other end of the barrel is the mold itself. They are typically made of steel, and must be precisely made. At the start of the cycle, the mold is closed. The injection process itself occurs when the entire screw is moved forward (but not rotated), forcing the molten material in the front section of the barrel through as series of pathways into the mold itself. After the molten resin is in place, it cools and solidifies, and then the mold is opened, and the finished parts are removed.

Injection molding is a fast process—typically in the order of seconds. It allows for detailed, high precision parts with high complexity. Since the entire process is done under pressure—very high pressure—the equipment and the molds must be made from high strength, high precision materials, and the mold halves must be held together with a high amount of force as well. The equipment and tooling for injection molding is expensive, but the speed of the process means the cost per part can be incredibly low. This method is used to make a wide variety of objects, from bottle caps, to disposable lighters, to musical instruments, to medical equipment.

Materials:	*Thermoplastics, some thermosets*
Equipment cost:	*High*
Tooling cost:	*High*
Part cost:	*Low–ultralow*

2.3.12 Reaction Injection Molding

Reaction Injection Molding (RIM) is a similar process to injection molding, except it uses a thermoset material instead of a thermoplastic. That means a curing process takes place in the mold, not just cooling, and instead of injecting a single (molten) raw material, two materials are mixed and injected into the mold, where they then undergo a chemical reaction. While the raw materials are more expensive, the lower material viscosity and lower injection pressure make RIM ideal for making large parts. It is often used in the auto industry for fenders and bumpers, and it is also ideal for large enclosures and housings.

Since the plastic is formed due to the chemical reaction, there is usually no need to heat the plastic before injecting it into the mold. This, along with the lower injection pressures, makes production equipment significantly less expensive than traditional injection molding.

Materials:	*Certain thermosets*
Equipment cost:	*Moderate*
Tooling cost:	*Moderate*
Part cost:	*Low*

2.3.13 Transfer Molding

Transfer molding is a bit of a hybrid between compression molding and injection molding. Like compression molding, it usually involves a thermoset material, and the mold is heated to allow the material to cure. However, instead of placing a slug of material in the mold, it is premixed in a separate chamber, called a pot. A piston then pushes on the pot, and the resin is transferred into the mold cavity. After heating and curing, the mold is opened, and the cured part is removed.

The molds are usually made of steel or cast iron, although they can be made of aluminum or other metals. The pressures involved are usually higher than with compression molding, but much lower than with the various injection molding processes. Transfer molding is often used to make electronic assemblies, where the electronic components are placed in the mold first, and then when the mixed resin is forced into the cavity, it flows over and around the electronic components, and they are then held securely in place when the material cures.

Materials:	*Thermosets*
Equipment cost:	*Moderate*
Tooling cost:	*Low–moderate*
Part cost:	*Moderate*

2.3.14 Structural Foam Molding

Structural foam molding is a variation of injection molding, with one major difference. Just prior to injection, an inert gas (such as nitrogen) is introduced to the molten resin along with a blowing agent. These then mix with the molten resin, and as the mixture enters the mold, the gas begins to expand and form a matrix of small bubbles—a foam. These small bubbles allow for even fill throughout the mold cavity, and as the molten resin cools and shrinks, the foam bubbles expand slightly, mostly in the inner core of the part.

Structural foam alleviates the requirement for very high injection pressures, and also for the very high clamping forces. This means the actual molds do not need the same level of structural integrity, so you can mold large parts using the structural foam process in smaller, less expensive molds, and with smaller, less expensive machines. It also allows for molding of very thick sections. Many large trash receptacles are made via the structural foam process (Figure 2.12).

Figure 2.12 A large trash receptacle, made using structural foam molding. WilleeColePhotography/Shutterstock.com.

Materials:	Many thermoplastics
Equipment cost:	Moderate
Tooling cost:	Moderate
Part cost:	Low–moderate

2.3.15 Expandable Foam Molding

Expandable foam molding is also a variation of injection molding, but differs from structural foam molding in that the foam is not added as a gas, but is instead created by a chemical reaction within the resin itself. It usually involves different kinds of materials, most of which are soft and pliable (and not structural).

Expandable foam also uses low pressure, so molds and equipment are usually inexpensive—even cheaper than with structural foam. Many automobile dashboards are made using this process (Figure 2.13).

Figure 2.13 Coffee cups, molded from expandable polystyrene foam. design56/Shutterstock.com.

Materials:	Some thermoplastics
Equipment cost:	Moderate
Tooling cost:	Low
Part cost:	Low–moderate

2.3.16 Blow Molding

Blow molding is a unique processing technology. It utilizes a number of phases, many of which are identical to the phases of injection molding, with the final phase similar to glass blowing. There are two versions of blow molding: extrusion blow molding and injection blow molding.

2.3.17 Extrusion Blow Molding

In extrusion blow molding, molten resin is extruded into a cylindrical shape called a parison. This parison hangs downward, and then two mold halves close around it. At the top and the bottom of the mold, the cylindrical shape is pinched together, leaving a hollow tube in the middle. A needle is inserted into this tube, and pressurized air is injected into tube, forcing the walls of the tube to stretch and expand outward, until they conform to the mold. Once the material cools and solidifies, the mold is opened and the part is removed.

Extrusion blow molding (and also rotational molding) enables the production of large hollow parts—sometimes very large. The size is limited only by the size of the machine, and of the mechanics of the molten resin as the parison is formed. It is only suitable for a limited number of materials, and secondary operations are usually needed to remove the excess material at the top and bottom, or to cut out unwanted sections.

Applications for extrusion blow molding are often similar to rotational molding. Extrusion blow molding is normally a faster process, but rotational molding can sometimes allow for thicker wall sections. Tools are usually steel, but can also be made of aluminum or other materials.

Materials:	*Some thermoplastics*
Equipment cost:	*Moderate*
Tooling cost:	*Moderate*
Part cost:	*Moderate–high*

2.3.18 Injection Blow Molding

The injection blow molding process involves two stages. In the first stage, a plastic part called a preform is made, using a traditional injection molding process. The preform is a precise shape, with precise wall thicknesses. This preform is then transferred to a blow molding station, where it is placed in another mold and held in place by specific features of the preform that mate with the blow mold cavity. Compressed air is then injected into the center of the part, forcing the unsupported walls of the preform tube to stretch and expand outward, until they hit the walls of the second mold. Once the material cools and solidifies, the mold is opened and the part is removed.

Injection blow molding allows for the production of hollow parts with some areas of the parts being very precise. It can be a very fast process. Applications for injection blow molding include plastic bottles, and plastic packaging with screw top lids (Figures 2.14 and 2.15).

Materials:	*Some thermoplastics*
Equipment cost:	*High*
Tooling cost:	*Moderate–high*
Part cost:	*Low–moderate*

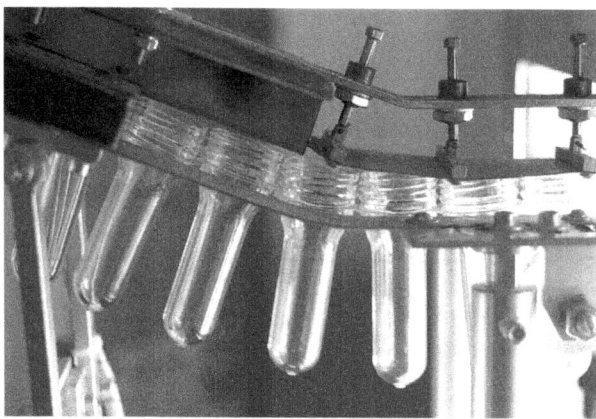

Figure 2.14 Blow molding preforms, being transferred to the blow molding station. DegtiarovaViktoriia/Shutterstock.com.

Figure 2.15 One half of a blow mold for a plastic bottle. GlebNsk/Shutterstock.com.

2.4 Process Comparison

While we have barely touched the surface of the myriad ways to process plastic, you hopefully now have a good idea of just how versatile these materials are. Sometimes the use of a specific process is justification alone for using plastic. Plastic is kind of like Bubba Blue's shrimp in the movie *Forrest Gump*. "Shrimp is the fruit of the sea," he said. "You can barbeque it, boil it, broil it, bake it, sauté it, … pan fry, deep fry, stir fry…" While I would not use any of these particular methods to process plastic, I like Bubba's attitude (Table 2.1).

Table 2.1 A List of Common Processing Technologies

Process	Materials	Production Details				Parts		
		Shapes	Quantity	Tooling Costs	Cycle Time	Complexity	Size	Cost
Vacuum forming	Amorphous thermoplastics, some semicrystalline	Discrete parts	100–10K	$	Minutes	Low	S–L	$–$$
Pressure forming	Amorphous thermoplastics, some semicrystalline	Discrete parts	100–10K	$	Minutes	Medium	S–L	$–$$
Extrusion	Most thermoplastics	Rods, tubes, films, sheets	1K–100M+	$–$$	Seconds	Low	S–L	¢–$$
Casting	Thermosets, some thermoplastics	Discrete parts	10–1K	$	Seconds	Medium	S–L	$$
Compression molding	Some thermosets	Discrete parts	1–100K	$–$$	Minutes	Medium	S–L	$$
Rotational molding	Some thermoplastics (mostly semicrystalline)	Discrete parts	1–10K	$$	Minutes	Medium	L–XL	$$

2: Why Use Plastic?

Injection molding	Most thermoplastics, some thermosets	Discrete parts	10K–100M+	$$–$$$	Seconds	High	XS–L	¢–$
Reaction injection molding (RIM)	Some thermosets	Discrete parts	1–100K	$$	Minutes	High	S–XL	$
Transfer molding	Some thermosets	Discrete parts	1–100K	$–$$	Minutes	High	S–L	$$
Structural foam molding	Some thermoplastics	Discrete parts	1–100K	$$	Seconds	High	S–XL	$–$$
Expandable foam Molding	Some thermoplastics	Discrete parts	1–100K	$	Minutes	Medium	S–XL	$–$$
Extrusion blow molding	Some thermoplastics	Discrete parts	1–100K	$$	Minutes	Medium	M–XL	$$–$$$
Injection blow molding	Some thermoplastics	Discrete parts	10K–100M+	$$–$$$	Seconds	High	S–M	$–$$

2.5 Plastics in Manufacturing

Plastic materials have a unique role in the world of manufacturing. They are primarily used as feedstock materials, in a similar way that flour is used in baking, or that rolled steel is used to make cars, or in how cement is used to make concrete. But plastic materials can also be easily processed to make intermediate products that are feedstock in their own right, including things like preformed sheets or tubes or other stock shapes, just as wood is processed to make lumber, and metals are processed to make pipes, and sand and clay and silica are processed to make bricks and tiles and window panes. The way in which someone looks at plastic materials often depends on their place within this manufacturing matrix, and on what end result they are trying to achieve in terms of using these materials. Are they trying to reduce cost? Or reduce weight? Or make an intermediate product that is easier to handle? Or mass produce an item in a way that is safer, cleaner, and more reliable?

Manufacturing, at its most basic level, is the process of making things. It is often a complex, highly mechanized process, involving factories, and capital investment, and machinery and people. Within this process, there is the specific act of creation, where something tangible is produced. We call this production. We often speak of something as being "in production"—meaning either that specific thing is in that act of, or that somewhere in the world, other things like that are currently being produced. Or we may speak of something as being "out of production"—meaning either the act of creation is complete for that specific thing, or that other things like that are no longer being produced anywhere.

In manufacturing, we typically think of production as a continuous activity, where multiples of something are produced in mass quantities, one after another after another. Or where multiples of something are produced in a finite number, more or less at the same time. However, production can also consist of making just one thing. The use of thermoplastics in manufacturing—including which plastic processing technologies are used—often depends on which method of production is involved.

2.5.1 Mass Production

Mass production, also called continuous production or flow production, involves the fabrication of a certain something in a specific way, in a specific shape, in a consistent manner. This something may be a stand-alone product—such as a Frisbee, or a rubber ducky, or a garbage can—or it may be a part—such as a screw, or a bolt, or a body panel—that is used to fabricate something else, or it may be an intermediate product that is used as a feedstock in the production of something else.

Thermoplastics are often used in the production of intermediate products. Just like iron ore can be converted into rolled steel, thermoplastic raw materials can be converted into solid shapes such as bars, sheet, plates, rods, and tubes. Some thermoplastics can be converted into a fiber form, which can be used as-is, or woven into cloth, or carpeting, other shapes. There is an entire industry dedicated to the fabrication and use of nonwoven thermoplastic fibers. Some thermoplastics can also be converted into a film form, which are often stored in rolls. And some thermoplastics can be converted into foams, which can be stored as sheets, in rolls, and sometimes as blocks. These intermediate products are often used as feedstock for other manufacturing operations, in either a mass production, batch production, or job production mode. We can take a block of thermoplastic material and then machine it to make something else—a custom part for a special machine, a window for the world's largest aquarium, or a liner for the socket of a human hip joint. We can take tubing made of another thermoplastic material and use it to make piping in our kitchen sink, or in a municipal sewer system, or in an aircraft engine. We can take a thermoplastic film and add a sticky substance to one side and make an adhesive tape. We can take a thermoplastic film and laminate it to another material and make sheathing for an electronic cable, or a bag that holds potato chips. We can take a thermoplastic foam and laminate it to something else and make a boogie board, or a cleaning pad, or a knee brace. The use of thermoplastics in the production of intermediate products is endless, and quite frankly well beyond the scope of this book.

The most demanding area of mass production is in the fabrication of standardized parts, particularly in the fabrication of complex parts. Thermoplastic materials are especially well suited for this area of mass production. Indeed, the material selection methods discussed in this book are dedicated to that endeavor. Some might even argue that thermoplastics are what makes this type of mass production possible. In this area of production, part costs and manufacturing efficiencies are just as important as performance and reliability and safety. Quite often, they also go hand in hand.

2.5.2 Batch Production

Batch production involves the fabrication of a finite number of things, usually within a very specific time frame. These things may be made all at once, as if one were making a large batch of cookies in a giant oven, or the fabrication might be done within a specific time interval, such as if you were printing a daily newspaper.

Batch production has certain advantages over mass production. The capital investment is usually less, and since it is done within a specific interval, one can often plan for that, and arrange materials, equipment, and people as needed. However, it also has its own set of challenges.

Thermoplastic materials are also used in batch production scenarios, albeit in a different manner. Instead of being used as raw materials, they are more likely to be used in an intermediate product, as a sheet that gets formed and cut, as a block that gets machined, as a film that gets converted to another shape, etc. In these scenarios, the material selection process is very similar to the process used in the fabrication of standardized parts for mass production. As such, the information presented in later chapters can also be used to evaluate the materials used in these intermediate products.

Interestingly enough, while the actual production of most plastic materials is done on a continuous basis, there are some plastic materials which are produced using a batch production method. Also, in many situations, thermoplastic materials are modified via a compounding process, which is a batch production method.

2.5.3 Job Production

Job production usually involves the fabrication of something on a very small scale—as measured in quantity. The something may be a collection of parts, or it may be a collection of things that are used to make something else. But normally, the phrase *job production* involves the fabrication of one very specific thing: a custom tailored suit, a house, a satellite, a space shuttle, a new World Trade Center.

As raw materials, thermoplastic materials are not often used in job production scenarios. However, just like in batch production, they are often used in intermediate products which are then modified for use a specific job. In this scenario, thermoplastic materials sometimes have unique advantages. Since the actual product that is being fabricating is unique, much of the cost in the fabrication is not in the cost of the material, it is in the cost for design and engineering and equipment and labor.

So, if you are building something unique, in a job production scenario you should select materials based on performance, feel, and end-use requirements. The actual material cost is going to be a minor component of the overall project cost.

2.5.4 Manufacturing Processes

Regardless of the production method used, there are a wide variety of processes that can be used in a manufacturing environment. In addition to the plastic processing technologies described earlier, there are technologies

2: Why Use Plastic?

for processing metals, ceramics, stone, clay, wood, and more. Some of these processes are focused on the raw materials, others on the fabrication of parts. There are also a wide variety of processing technologies for the finishing and assembling of parts—on either on individual basis, or on a system level.

In the early days of the Industrial Revolution, when people first started making parts and systems on a mass scale, it was often enough to rely on the cost and precision advantages afforded by a factory-based production system. In today's world, with an ever increasingly competitive world economy, that is not enough. We need to design parts and products in such a way that we eliminate unnecessary processes.

- Perhaps you can design a part—made out of a specific thermoplastic—that replaces the need for a part made of steel, and in the process eliminates the need for chrome plating—and the use of sulfuric acid and hexavalent chromium.
- Perhaps you can design a plastic part with molded-in threads, and eliminate the need for machining a metal part, and for the cleanup (and disposal) of the metal chips and the solvents and lubricants used in the machining process.
- Perhaps you can design a part—made out of a specific thermoplastic that is stiff and strong—that eliminates the need for machining and heat treating a metal part, while eliminating the capital expenditure for the equipment required for that process, and for the energy costs involved.
- Perhaps you can design a system made out of plastic parts that can be snapped together instead of bunch of metal parts that need to be welded in place (and then painted or otherwise coated to prevent corrosion).

In these types of situations—and in many others—thermoplastic materials can fulfill a unique role. But this concept of eliminating unnecessary processes is not unique to the manufacturing processes used to make products from thermoplastics. As an example, the housings in many products made by Apple are machined out of solid billets of aluminum. At first glance, this seems incredibly expensive. One not only has to pay for the entire block of material, but also for all of the machining operations, as well as the costs to clean up and dispose of all the metal chips. However, when done well, this type of manufacturing process allows for exceptionally high precision. Assembly of the remaining components can be carefully controlled, and the fit, form, and function of the final product has set a new standard for consumer satisfaction. In many ways, this is the ultimate goal of mass production.

3D printing is another example of eliminating unnecessary processes in a manufacturing environment. In 3D printing, the printer takes a computer generated model—one that exists only in virtual reality—and uses it to create a real-life object, one that is precise and accurate and true to the virtual model. While the process has been around for decades, the technology is advancing at a rapid rate, to the point where we are now able to 3D print a number of different materials, including a wide variety of thermoplastics, certain metals, concrete, even human bone, and living human tissue. While this technology is probably best suited for batch production methods, the day will soon come when we will be able to 3D print fully functional parts—in any material we want, in any quantity we want, without the need for secondary processes, investment in tooling, or waste of material.

Truth be told, manufacturing success stories are easy to achieve. All it takes is good design, intelligent engineering, and a business structure that allows the design team to explore creative manufacturing solutions (although it can be difficult to find a manufacturing company that has one of those attributes, let alone all three).

One item of note: thermoplastic materials are rarely used in the tools used in production. Yes, there are housings and jigs and fixtures made from thermoplastics, but the cutting tools themselves, the saw blades and drill bits and milling cutters, are still made from high-strength steels and other specialty metals. The same is true for production equipment: the cranes and forklifts and trucks and dollies and overhead tracks. Some of the components in this equipment are made from thermoplastics, but the heart of the equipment itself is made from metal. The reasons for this are discussed in later chapters.

2.6 Advantages of Thermoplastics

Thermoplastic materials offer a wide range of advantages in performance, processing and manufacturing options, safety, and cost.

2.6.1 Performance

In evaluating performance, we often focus on the ultimate, the absolute best performance: the world's fastest man, the tallest building, the highest mountain, the deepest ocean. Like all materials, thermoplastics have a combination of physical, electrical, and thermal properties, however, very rarely will a thermoplastic material have the absolute best performance in any single measured property. Instead, what often sets them apart is a combination of properties that taken together, provide better performance than other materials. Some examples include:

- stiffness versus weight—when compared to metals, thermoplastics have very low densities: about 1/3–1/2 the density of aluminum, and 1/8–1/10 the density of steel. So even though their ultimate stiffness may be lower, the stiffness-to-weight ratio can be higher, allowing for lighter weight structures.
- strength + corrosion resistance—when compared to metals, most thermoplastics have a lower ultimate strength. However, if you add in a requirement to withstand ocean water, or caustic cleaning agents, a thermoplastic may have better performance.
- heat resistance + chemical resistance—perhaps you want to boil something in sulfuric acid (for whatever reason). Not a lot of materials that can handle that, but there are thermoplastics that can.

2.6.2 Processing Options

As discussed earlier, there are numerous processing technologies that are suitable for use with thermoplastic materials, which allows for fabrication of plastic components in a wide variety of shapes and forms. These technologies allow thermoplastic materials to be formed not only into solid shapes, but woven and nonwoven fibers, films, tapes, and foams.

2.6.3 Near-Net-Shape Manufacturing

Many of these processing technologies allow for part consolidation, where features and functions of the final product, which previously required one to fabricate two or more discrete parts, can be achieved in a single plastic part.

Some of these processing technologies, such as injection molding, also have the ability to create near-net-shape parts. These processes often result in a part that is close to its final (net) shape, in a single operation. They offer high precision, and are accurate and repeatable. They also allow for fabrication of complex shapes.

2.6.4 Safety

Believe it or not, thermoplastic materials have set new standards for safety, not just in terms of product safety, but also in terms of manufacturing safety, and environmental safety.

Thermoplastic materials are used to preserve our food from spoilage and contamination. They are used to store our medications, and to prevent the spread of infection and disease. They are used to make bulletproof vests, fireproof clothing, and shatterproof eyeglasses. They are even used as guards and shields over other "better" materials as extra precaution [1].

Products are often made out of thermoplastics specifically to prevent injury to people when they use and abuse the product in an unsafe manner. (Any good designer knows how difficult it can be to design a product that cannot hurt anyone, even when used by idiot, because some idiots can be rather clever.)

The manufacturing of plastics and rubber products is among the safest of all manufacturing industries. Injuries and fatalities among workers in this industry are significantly lower than for logging and mining operations, or for building construction, or for fabrication and processing of iron and steel. While logging takes the top of the list among the deadliest of jobs, others include fishing, piloting aircraft, garbage collection and processing of recyclables, truck driving and transportation, and agricultural work [4,5]. And recently, a coalition of plastics professionals and unions founded Safety in Manufacturing Plastics and Composites (SIMPL), an initiative whose aim is to reduce major injuries in the workplace and improve health and safety standards [2,3].

Finally, thermoplastic materials are inherently friendly to the environment. Not only are they inherently recyclable, their production is earth friendly. There is no mining of heavy metals, no clear cutting of forests, no runoff of silt, and no acid rain. Their transportation and use requires far less energy than other materials. Yes, there are issues with recycling and reuse and end-of-life planning—as with all materials used in manufacturing—but these issues can be solved.

2.6.5 Cost

One of the major advantages that thermoplastics have over other materials is in the area of cost. Not only are the raw materials costs significantly lower than with other materials, but their processing costs are often lower as well. It is often possible to fabricate a part or intermediate product out of a thermoplastic in fewer processing steps than what would be needed to fabricate a similar part or product in another material. Fewer steps equal lower costs. And more often than not, secondary finishing operations, cleaning, prepping, painting, anodizing, etc., can be eliminated, resulting in even lower costs. Near-net-shape manufacturing typically results in low piece part cost. And the density of thermoplastics means lighter weight parts, so shipping costs are lower.

2.7 Disadvantages of Thermoplastics

Thermoplastic materials also have some disadvantages. Some of these, such as heat resistance, are easy to understand. Others require some explanation.

2.7.1 Heat Resistance

All thermoplastics have a melting point. While there are plastic materials with exceptionally high melting temperatures and/or exceptionally high in-use temperatures, these temperatures are typically lower than for metals, glass, and ceramics. As a result, thermoplastic materials would not perform as well as these materials in high heat applications. While there are some specialty thermoplastics that can withstand very high temperatures, I do not think you are going to see thermoplastic light bulbs any time soon.

2.7.2 Time-Dependent Behavior

Plastic materials—when compared to other materials—have a unique relationship to time. What do I mean by that?

If we think of speed, that is velocity—as a measurement of distance traveled over a given interval of time—we can then describe speed as a distance–time relationship. If we talk about the rate of an event—as in how frequently something happens in a given interval of time—we could describe rate as a frequency–time relationship. Plastic materials are quite peculiar when it comes to their performance in these kinds of time relationships.

So how are plastics materials peculiar? First of all, the mechanical properties of most plastic materials are highly rate sensitive. As an example, think of silly putty. Silly putty is a children's toy, and consists of a small blob of material encased in a plastic egg. If you take the blob in your fingers, and pull it slowly, it will stretch and elongate, much the same as blowing bubbles out of bubble gum. But if you pull it quickly, the blob will

Figure 2.16 Silly putty being pulled.

snap in half, with a clean break. Same material but two different behaviors based on the speed at which the material is pulled apart. This is rate-sensitive behavior (Figure 2.16). We can often account for this kind of behavior through proper design, and/or proper material selection, but the funny thing is, most property data for thermoplastic materials are generated from data gathered under controlled conditions—at very specific rates of loading. If the rate of loading in your application is different from the test scenario—what do you do?

Secondly, most plastic materials are susceptible to creep. Creep is the tendency of a material to slowly deform under the long-term influence of applied stress. While many materials are susceptible to creep, it is often in relationship to the difference between the melting temperature of the material and the in-use temperature of the application. Since the melting point of most plastics is well below the melting point of other materials (metals, ceramics, etc.), creep at room temperature is more of an issue for plastic materials.

Finally, there is the issue of long-term use. On a macroscale—years, decades, centuries, millennia—we do not have a whole lot of data on the long-term behavior of plastics. There are some long-term data (based on aging studies, and analysis of actual parts in use for long periods of time) on some of the more common materials. And we also know that some plastic materials degrade upon exposure to UV rays (as found in sunlight), others upon exposure to heat, or chemicals, etc. Much of these kind of data are available—for short-term exposure. But for many materials, especially newer materials, the data for long-term use just do not exist.

2.7.3 Temperature Variations

Another disadvantage with thermoplastics has to do with how they respond to temperature variations. Almost all materials expand upon heating, and contract upon cooling. One of the challenges with thermoplastics is that their rate of expansion is 5–10 times greater than that of metals (or concrete or cement), and 20–30 times greater than with wood or clay. This can make fitting parts together extremely difficult, especially when you are making large parts out of different materials that are used over a wide temperature range.

Most thermoplastic materials also have a significant change in properties as the temperature changes. They may become very flexible (and often weaker) at high temperatures, and stiff and brittle at low temperatures.

2.7.4 Structural Inconsistencies

On a material science level, thermoplastics also have different behaviors than many other materials. For starters, most thermoplastics are anisotropic; that is, they have different properties when measured in different directions. They also have different behavior in compression than they do in tension. And their mechanical behavior is nonlinear, in that their behavior does not follow the traditional linear stress–strain relationship seen in metals. This means the classic engineering equations for structural calculations are not always accurate.

On a part level, molded plastic parts also often have significant variations in the structure of the plastic material in various areas of the parts. There may be areas of high molded-in stress, variations in crystallinity, variations in density, variations in how the polymer molecules are aligned, variations in how additives and reinforcing agents are dispersed, etc.

On the other hand, many other materials have structural inconsistencies as well. All wood has a grain, and the properties of a given piece of lumber will vary depending on whether you measure with the grain, through the grain, or across the grain. They will also vary depending on what part of the tree the board was cut from. Cloth has directional variations due to the weft and warp of the fibers used. And many rocks exhibit directional fissility, which is the tendency to split into long flat pieces. Other rocks such as flint, flake into sharp blades—which is what made them so good for making arrow heads in the first place.

Fortunately, a good plastics engineer knows how to account for these structural inconsistencies.

> *A foolish consistency is the hobgoblin of little minds, adored by little statesmen and philosophers and divines.*
> **Ralph Waldo Emerson, American poet**

2.7.5 Repairing Broken Parts

Broken plastic parts can be difficult to repair. Sometimes, it is simply not possible. This is due to not only the behavior of thermoplastic materials, but also the processing methods used. Thermoplastic parts are formed under a combination of heat and pressure, and for most forming processes, the pressure is substantial, and requires hardened steel molds and special machinery. Most end-users would not have access to this kind of equipment, and even if they did, they probably would not have a mold in the shape of the broken feature. One might be able to use a soldering iron or a heat gun to melt the material in the broken part, but in order to reform the molten plastic in situ one needs pressure, and a mold.

Sometimes broken parts can be repaired using high-strength adhesives. However, more likely than not the properties of the adhesive will be different than those of the base material, so while the repair might "work" for a period of time, the performance of the repaired part will be different than the performance of the original part.

In some cases, it may be possible to replace a broken part, provided one can disassemble the item to isolate that one broken part. Some companies go to great lengths to provide replacement parts for their products, but in many industries the cost cannot be justified, and the user has to either throw away the product, or jury-rig a solution. Advances in 3D printing may soon allow us to fabricate replacement parts on the spot, as they are needed, using a material that is nearly identical to the original production material.

In many cases, the product may be designed or manufactured in a way that precludes removal and replacement of a broken part. This is true in many industries, and not just with products made from thermoplastics. Many automobile parts for instance, are built as subsystems, and when a part fails in that subsystem, the entire subsystem must be replaced.

The problem of repairing broken parts is not unique to thermoplastics. Indeed, if you had a gear made of cast iron, and one of the gear teeth broke off, how would you repair it? Or if you had a large bell (Figure 2.17), cast from a mix of copper and tin, and shortly after it was cast it developed a crack, how would you repair it?

Figure 2.17 The Liberty Bell. vectorkat/Shutterstock.com.

2.7.6 Perception

Another disadvantage of thermoplastics is how they are perceived by many people. Most of us who work these materials know their strengths and weaknesses, and we design products in such a way as to utilize thermoplastics to their advantage.

But to the vast majority of people, plastic materials are often perceived as being cheap—even worthless. They are perceived as weak, flimsy, or otherwise inadequate when compared to other "better" materials. A cyclist will take a mountain bike into the woods, and then after crashing into a tree, he will look at a broken aluminum brake lever and go, *"Wow, I must have really abused that thing."* That same cyclist will take his bike into the woods, and then after he hits a tree stump with his derailleur, will look down at a broken plastic component, and say, *"Cheap plastic."* He will then brush off his polyester and spandex jersey, put his polycarbonate helmet on, put his nylon backpack on, and head back to his car, loaded with plastic components, and then drive home, muttering to himself and complaining to anyone who will listen about these pathetic plastic materials.

Perception is Reality
Lee Atwater, Political consultant

2.7.7 Human Behavior

By far the biggest disadvantage of using thermoplastics has to do with human behavior. People often say one thing about plastics, but their actions around plastics tell a different story.

On an individual level, people may describe plastics as being cheap and inferior to other materials, yet they use products made of plastic materials on a daily basis, and many of those products outperform products made with other materials. They will take their most precious possessions, and then put them in water-tight plastic storage cases, to preserve and protect something they value. They will take live-saving medication, and put it in tamperproof, waterproof, UV resistant plastic bottles. And when they are done with any of these things, they will take those plastic items and throw them away. *Cheap plastic*. Then they will turn around and go shopping and do some price comparisons and buy another product at the lowest possible cost, regardless of the materials it is made of, or the quality of its construction. Then, when a plastic something in that product fails, they throw the whole thing away. *Cheap plastic*.

On a societal level, we have a love affair with garbage, and a lot of that garbage is plastic. People may say they do not like to waste things, and that garbage is nasty and unsightly and we need to reduce it. And people say we should reduce, reuse, recycle things, and yet, every year, on a per capita basis people are creating more and more garbage. As a species, we are not walking the talk.

On an industry level, the plastics industry is woefully lacking in the development of a cradle-to-grave, zero-waste, fully closed-loop recycling solutions. Oh sure, there are some plastic recycling centers here and there, where you can take empty soda bottles and some packaging. But most people need incentives to recycle, and in many places, people are required to go out of their way to recycle even the most environmentally friendly materials. The plastics industry has failed to create any cohesive strategy to enable recycling on a mass scale.

An associated problem is a general lack of recycling infrastructure. Most thermoplastics are inherently recyclable, yet many areas of the world—including large parts of the United States—do not have any systems in place for efficient recycling. Where systems are in place, it is still often easier to throw things away, and on a societal level it is even encouraged. We teach kids to clean their room, pick up the trash, and then take out the garbage. Why are not we teaching kids about a complete closed-loop system, where everything in the system is connected to everything else, a waste product from one area is a feedstock in another, and everything has a use, and a reuse, along with planning for end-of-use, and the entire system is self-perpetuating?

On a political level, the ignorance and beliefs about thermoplastics is mind-blowing. We have politicians who are convinced that plastics are evil and bad for the environment and their use needs to be controlled and curtailed. Instead of passing laws that ban the use of plastic grocery bags, why not pass a law that bans an individual—any individual—from throwing away an item that can be recycled? Why not pass laws that enable and support the development of a cradle-to-grave, zero-waste, fully closed-loop recycling infrastructure? One of the purposes of government is to ensure the welfare of its people. Perhaps it is time we start electing politicians who understand the role that plastics have in ensuring that welfare.

2.8 The Uniqueness of Thermoplastics

As we discussed in Chapter 1, humans have used different materials throughout history. And while the usage of any given material may have changed over time, the fact is that we still make things out of wood, stones and shells, concrete and clay, rocks and bricks. Sometimes materials are replaced because they have become scarce, or too expensive; but for the most part, we use a certain material because of its unique qualities. The same is true for thermoplastics.

Thermoplastics offer a unique combination of performance, manufacturing options, safety, and cost. As raw materials, they have the added advantage of being available almost anywhere in the world. Furthermore, as synthetic man-made materials, we have the ability to create new derivative materials, by using blends of existing materials, or by modifying their chemistry, or in the use of custom additives. This ability to create new and unique materials is unprecedented in human history. Who knows what the next great super plastic will be?

Thermoplastics also have a unique characteristic in how they sound, how they feel, and how they look. We often make associations based on how different types of materials perform in these areas—and we have responses based on those associations. Thermoplastic materials offer an opportunity to affect those responses. This will be discussed in greater detail in a later chapter.

2.9 And the Answer Is...

So, after all of that, let us ask the question again. *Why Use Plastic?*

Or, to paraphrase the question slightly, "Why should I use plastic materials in the design of my new system?"

There are times when the answer is—*You should not.* There are lots of delightful things in this world that are made from other materials, and rightly so. But there are times when you should make a certain something out of plastic, for one or more of the reasons discussed in this chapter. But the funny thing is, at least in the world of mass of production, 9 times out of 10 the logic behind the answer goes something like this:

We can make this thing out of a thermoplastic—safely, reliably, and consistently—at a lower overall cost than with any other material.

References

[1] America's Plastics Makers, Plastics: Safety and Consumer Benefits, American Chemistry Council, 2014.
[2] Health and Safety Executive, The SIMPL Initiative, 2012.
[3] Health and Safety Executive, Safety in Manufacturing Plastics 2011–2014 (SIMPL), 2014.
[4] Jacquelyn Smith, America's Ten Deadliest Jobs, Forbes Magazine, August 22, 2013.
[5] United States Department of Labor, Bureau of Labor Statistics, 2014. http://www.dol.gov.

3 Understanding Thermoplastics

3.1 Introduction

Making something out of a thermoplastic material is very different than making something out of metal, glass, ceramic, or even wood. Not only are you constrained (or perhaps liberated) by the plastic processing methods involved, but there are often a multitude of other constraints, many of which seem to contradict each other.

The materials themselves are also different—not just in terms of chemical makeup, but in mechanical behavior as well. They bend and deform in a different manner, respond to their surrounding environment in a different manner, are affected by temperature and time in a different manner. Working with thermoplastics requires an entirely different thought process. To paraphrase a famous comedian, "Learning a new language is so hard. They have a different word for everything."

Many designers and engineers begin the process by collecting data sheets on different thermoplastic materials, and start comparing the numbers for various properties, tensile strength, flexural modulus, heat deflection temperature, etc. in an attempt to find "the best material" for their project. Personally, I find this approach counterproductive. It is almost like trying to analyze your monthly cash flow by seeing how much money you have in your checking account at 3:17 PM on a Tuesday. That may be a useful piece of information (especially if you want to withdraw a certain amount of money at 8:00 AM Wednesday), but it is often not relevant to the big picture.

Indeed, ASM International (formerly known as the American Society for Materials), provides this caution on the use of property data sheets:

> The designer who has access to standard data sheet values only must use them cautiously as a source of design information. Many of the tests reported on the sheets come under the jurisdiction of the American Society for Testing and Materials (ASTM), and the properties are intended to characterize a material, not necessarily to provide design information.[1]

To use thermoplastic materials effectively, one must understand not only their unique behavior, but how that behavior is affected over a broad

---

1. Engineered Materials Handbook, Volume 2, Engineering Plastics. ASM International.

span of conditions—not just the environmental conditions in the end-use application, but also the processing conditions the resin itself has gone through. To help develop that understanding, we need to start with a review of some basic fundamentals of materials science. More than just discussing some theoretical concepts, we need to appreciate the interconnected nature of material structure, property data, processing, and end-use performance. And in the case of thermoplastics, we also need to review some of the fundamentals of polymer science.

This chapter is not intended to be a substitute for a comprehensive training program in materials science or plastics engineering. Rather, it is intended to highlight some key issues, in order to bring some clarity to the complex world of thermoplastic material selection.

3.2 Materials Science

Materials science is an interdisciplinary field, involving aspects of physics, chemistry, mathematics, even thermodynamics. It has its origin in the use and processing of metals. One of the fundamentals of material science is that the physical properties of a material are related to its atomic structure. The field has been helped, and been driven by, the development of new materials, including plastics.

3.2.1 Strength of Materials

Physical properties of a material include mechanical properties, electrical properties, thermal properties, and more. The mechanical properties of a material deal with how it responds to physical forces. Compared to other materials, thermoplastics respond to force in unique ways.

The most basic physical property has to due with the concept of strength. In engineering terms, the strength of a material is its ability to withstand an applied force without failure. Force is a term from classical mechanics, and is a vector, meaning both magnitude and direction are involved. In materials science, strength is quantified on a force per unit area basis. Also known as load, or stress, the forces can be applied in tension (the material is being pulled), compression (the material is being pushed), bending, torsion (twisting), or shear (loads in the same plane as a cross-section of the material) (Figure 3.1).

Regardless of the type of load, as the load is applied, the material will alter its shape, or deform. This deformation may be temporary, that is the material will return to its original shape when the load is removed, or

Figure 3.1 Types of loads: compression, tension, bending, torsion, shear.

it may be permanent. Temporary deformations are referred to as elastic deformation, and permanent deformations are referred to as plastic deformation (or failure, depending on their severity). Most materials, including steel, can experience both elastic and plastic deformation. The deformation itself is measured on a percentage basis, and is known as strain. While the deformation of different-sized objects will be different, the stress versus strain behavior of a given material is predictable and repeatable.

Typically, strength measurements are made with one specific type of load, applied in one specific direction. Testing is done under controlled conditions, where the load is gradually applied, and measurements of stress and strain are recorded continuously. The resulting data is then plotted, with stress on the vertical axis, and strain on the horizontal axis. These plots are known as stress–strain curves (Figure 3.2).

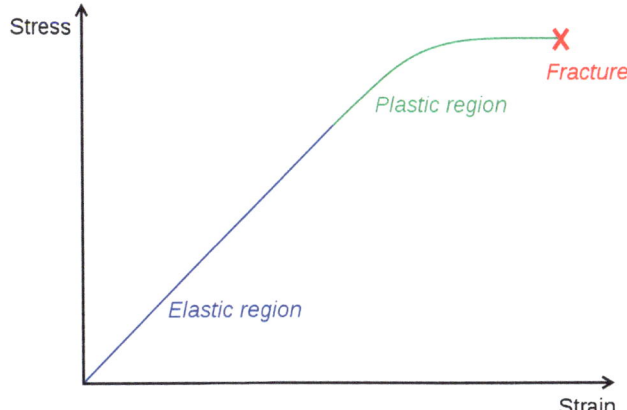

Figure 3.2 Generic stress–strain curve for a ductile material.

The exact shape of the stress–strain curve will depend on the material, the rate of loading (how quickly and/or slowly the load is increased), and the temperature. The ultimate failure of the material can be abrupt (and often quite dramatic), where it simply snaps in half. This type of failure is called a brittle failure. The failure can also be slow and gradual, where the material begins to stretch and thin out, and eventually breaks. Or the failure can be a mixture of the two. The ratios of the elastic to plastic regions can also vary. But each material has a unique stress–strain curve, almost like a fingerprint.

In the elastic region of the plot, most materials deform in a linear manner, that is, for any given amount of stress (commonly denoted by the Greek letter σ), there is a given amount of strain (commonly denoted by the Greek letter ε), and the relationship between them is constant. This constant is the Elastic Modulus, E, also known as Young's Modulus, where:

$$E = \sigma/\varepsilon$$

For any given stress σ_1, there is a corresponding strain ε_1. For a linear material, since E is a constant, if $\sigma_2 = 2 \times \sigma_1$, then $\varepsilon_2 = 2 \times \varepsilon_1$.

It is important to note that the elastic modulus can be measured in any of the loading situations described earlier, tension, compression, shear, torsion, or bending. For materials that are homogeneous (uniform and consistent) as well as isotropic (having the same properties in all directions), there are simple equations that correlate the elastic moduli of the various loading conditions. Isotropic materials include metals, glass, ceramics, sand, many rocks and stones, etc. One could even argue that the mathematics used in materials science is based on the premise that materials are isotropic.

In the plastic region of the plot, the mathematics of the stress–strain relationship become much more complicated, and is well beyond the scope of this book. However, the behavior of a material in this region, how it stretches and deforms into a new shape on its path to failure, is in many ways what makes each material unique.

3.2.2 Anisotropic Behavior

The first thing to be aware of with thermoplastic materials is that they are anisotropic, that is their properties vary depending on the direction that is being measured. There are many reasons for this, one of which has to do with how the polymer molecules are distributed within the material, and whether they are organized and aligned or randomly distributed. The distribution depends not only on the characteristics of the specific thermoplastic, but also on the processing technologies involved. Many processes will result in alignment of the polymer molecules in a specific direction, and the properties as measured in that direction will be markedly different from the properties measured across that direction. This is not always a bad thing; there are times when these orientation affects are intentional, and we can take advantage of them.

Most plastic processing technologies have a practical limit to the maximum thickness of the final product. The resulting shape—although three dimensional—will have distinct length, width, and thickness parameters, and the properties in the thickness direction will be distinctly different than in the length or width direction. In many cases, the properties will vary at various points *through* the thickness, with one value at the bottom surface, another in the middle (or core), and yet another at the top surface.

Furthermore, most thermoplastics have distinctly different properties in tension and compression—even when measured in the same direction. Most thermoplastics resist compressive loads much better than tensile loads; the deflections are smaller, and the ultimate loads before failure are higher (this is also true for concrete). As a result, the elastic modulus in tension is often different than the elastic modulus in compression. Sadly, many engineers and designers fail to consider this. In real life, most end-use applications involve a combination of loading conditions; some tension here, some compression there, some shear here, some bending there. Most property data sheets provide some specific measurements of bending behavior, in the form of Flexural Modulus (the elastic modulus of a beam in bending), and Flexural Strength (the ultimate strength of a beam in bending) (Figure 3.3).

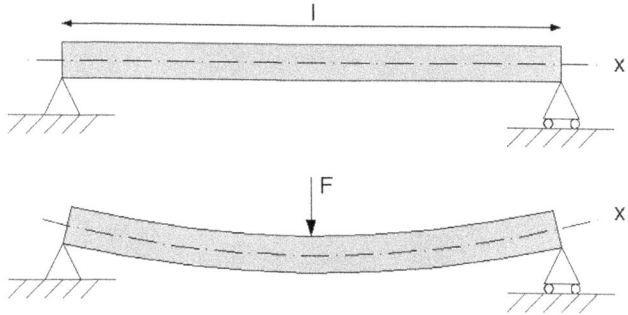

Figure 3.3 A simply supported beam before and after the application of a force, F.

3.2.3 Nonlinear Behavior

This second thing to be aware of with thermoplastic materials is that their behavior in the elastic region is nonlinear. Unlike linear materials, where the relationship between stress and strain is constant, in a nonlinear material the amount of additional strain for an additional amount of stress varies—depending where you are on the stress–strain curve. And if you do not know where you are on that stress–strain curve, it means that predicting the incremental strain is impossible. It is like giving someone who lost a map, and then telling them to find their way home—*without telling them where they currently are*.

The significance of this is that if you use standard engineering equations for mechanical behavior to calculate in-use performance of a part made from a thermoplastic material—almost all of which are based on linear mathematics—the results will only be a theoretical approximation.

3.2.4 Stiffness

Another basic physical property has to due with the concept of stiffness. In engineering terms, stiffness refers to the ability of object to withstand an applied force without deformation. Stiffness is dependent on both the structure of the object, and the elastic modulus of the material(s) being used. Stiffness can be measured in bending, in twisting (torsion), or in a single direction (tension or compression in one axis) (Figure 3.4).

In common language, the term stiffness is often used interchangeably with elastic modulus; materials with a high modulus are described as

Figure 3.4 One aspect of stiffness is the ability to withstand bending. PeJo/Shutterstock.com.

3: Understanding Thermoplastics

being stiff, and materials with a low modulus are described as being flexible. While technically this is an improper use of the term, one needs to understand that it is often used this way.

While stiffness and strength are related, they are not dependent on each other. In other words, a material can have high stiffness and low strength, or low stiffness and high strength, or a combination in between.

3.2.5 Toughness

Another basic physical property has to due with the concept of toughness. In engineering terms, the toughness of a material is its ability to withstand sudden impact (Figure 3.5).

Toughness can be difficult to quantify. How sudden is sudden? As described earlier, all thermoplastics are rate sensitive, so the measured value of an impact at one rate will be distinctly different than the value of an impact at another rate. Then there is the question of the kind of impact. Is it a shear load?; A tensile load?; A compressive load?; A

Figure 3.5 Toughness is the ability to withstand sudden impact. JohnTTakai/Shutterstock.com.

twisting load?; A combination load?; Is it a gouge, scrape, poke, slam, kick, or a bite? The response of the material will be different for all of these. Fortunately, there are numerous standard test methods in the plastic industry, and any or all of them can be used to quantify various aspects of toughness.

A simple working definition of toughness is this: the ability to withstand abuse. Or, as a former colleague once said, "Toughness is whatever property it is that is lacking in this part that just broke"[2].

3.3 Polymer Science

Polymer science is a specialized field of materials science that focuses on the study of polymers. In addition to the unique characteristics of thermoplastics described earlier (anisotropic behavior, nonlinear elastic response, etc.), polymer science also has a unique language (Figure 3.6). Just like the word *plastic* has many meanings, many of the words used in polymer science have many meanings. In this section, we will review some of the more important terms, in order to provide a platform for subsequent discussion.

Monomers and polymers—monomers are the basic building blocks of plastics. They are small molecules, typically consisting of hydrocarbons. Monomers then form chains or strings with other monomers (of the same or various types depending on the desired properties) in a process called *polymerization*. The resulting chain of monomers is a polymer. The characteristics of the monomer along with the properties of the resulting polymer chain yield the distinct qualities of a particular type of plastic. While the term polymer is often associated with plastic, polymer structures occur in nature as well, like glucose and natural rubber (Figures 3.7 and 3.8).

When a single type of monomer is used, like ethylene, the resulting polymer is a *homopolymer*. When two different types of monomers

---

2. A subtle paraphrase of a famous quote—"I'll know it when I see it." This 'quote' is often loosely attributed to Potter Stewart, former Associate Justice of the Supreme Court of the United States. The complete actual quote is as follows: "I shall not today attempt further to define the kinds of material I understand to be embraced within that shorthand description [hard-core pornography]; and perhaps I could never succeed in intelligibly doing so. But I know it when I see it, and the motion picture involved in this case is not that."

Figure 3.6 The world of polymer chemistry has a language unlike any other. GorbashVarvara/Shutterstock.com.

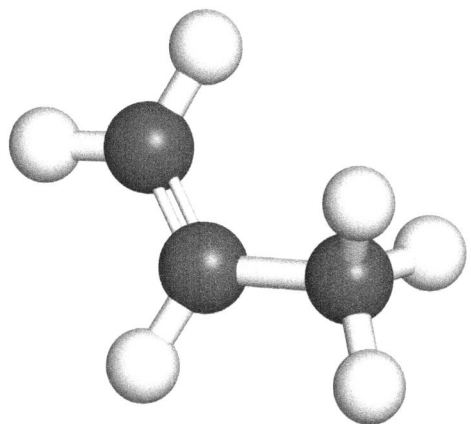

Figure 3.7 Propylene monomer. molekuul.be/Shutterstock.com.

Figure 3.8 Polypropylene polymer. molekuul.be/Shutterstock.com.

are polymerized into a single polymer chain, the resulting polymer is a *copolymer*. If three types of monomers are polymerized, it is a *terpolymer*.

Resin—resin is a term used within the plastics industry to describe plastic materials. It is often used interchangeably with the word plastic, but it specifically describes the material itself, whereas the word plastic can describe not only the material, but also engineering behavior (plastics deformation, etc.).

Chemical name—all polymers have a specific scientific name that is based upon the chemistry of the polymer itself. It usually includes the name of the primary monomer used, and may also refer to the chemical bonds in the polymer chain.

Chemical family—most polymers are also part of a chemical family that is, there are other polymers with a similar polymer structure, but with minor deviations in the monomers used, or in the bonds. Polymers within a given chemical family typically have very similar properties.

Common name—most polymers also have a common name, usually one that is simple and easy to pronounce. Unfortunately, common names sometimes vary, depending on where in the world one is.

Amorphous—the word amorphous comes from the Greek work *amorphos*, which literally means without shape (*a-* meaning without + *morphos* meaning shape). In polymer science, amorphous refers to a material where the polymer chains arrange themselves in a random, haphazard manner. They are distinctly different from crystalline materials, where the molecules are oriented in a regular, repeating pattern (Figure 3.9).

Crystalline—crystallinity is a term used to describe a specific type of ordered structure in a solid material. In a crystalline material, the molecules are oriented in a regular, repeating pattern. Most metals are crystalline, as is quartz, which forms a very distinct crystalline structure. In the world of thermoplastics, there are many materials, which have *some* degree of crystalline structure, but only a very few that are 100% crystalline. As a result, most of these materials are described as being *semicrystalline* thermoplastics (Figure 3.10).

Crystallization—a term used to describe the process as a material develops its crystalline structure. The process is unique for each polymer, and can be affected by a number of factors, including wall thicknesses in the part, nucleating agents in the material, the heat transfer dynamics

Figure 3.9 A plate of spaghetti, with the strains of spaghetti arranged in a random, haphazard manner. Viktor1/Shutterstock.com.

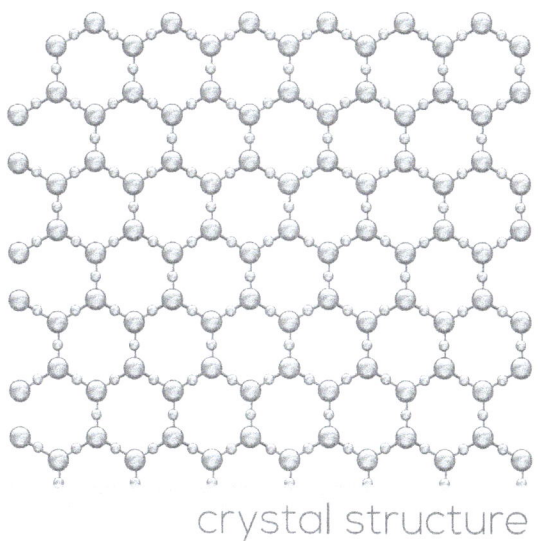

Figure 3.10 A molecular model of a crystalline structure, with the atoms arranged in a regular, repeating pattern. kirate/shutterstock.com.

of the molding process etc. The degree of crystallinity in the final part is a function of both the crystallization process and the characteristics of the polymer.

Glass transition temperature—commonly designated as T_g, the glass transition temperature represents an important phenomenon in polymer science. Below this temperature, the amorphous polymer molecules are in a rigid state. While they are randomly organized, they are also fixed, and the structure is hard and often brittle—just like glass. Above this temperature, the amorphous polymer molecules transit into a flexible state, resulting in a softer, more malleable material. While the T_g is often given as a specific temperature, more often than not it is a range of temperatures—below that range the material is hard, above that range the material is soft (very soft). Semicrystalline materials will also have a T_g, but it usually only affects the amorphous (noncrystalline) areas of the material, so it may not have a significant role in the overall performance of the material.

Melting temperature—commonly designated as T_m, the melting temperature of a polymer is defined as the temperature at which the crystalline structure breaks down, and undergoes a phase change into an amorphous state. The T_m is predictable and reversible—below that temperature the material has a crystalline structure, above that temperature the material is in an amorphous state. By definition, amorphous materials do not have a distinct melting temperature.

Melt temperature—melt temperature is the actual, measured temperature of a molten plastic. Most thermoplastics are processed under a combination of heat and pressure, and as they soften with the application of heat, they can also be formed with the application of pressure. Semicrystalline materials typically need to be processed with a melt temperature above the T_m, whereas amorphous materials can typically be processed over a range of temperatures and pressures (as long as the temperature is above T_g). Most resin suppliers provide recommendations for melt temperature for the resins they sell.

Molecular weight—molecular weight refers to the mass of a specific polymer chain. Within a given polymer, a higher molecular weight corresponds to a longer polymer chain, and usually a longer polymer chain means the molecule has better properties (higher strength, etc.). However, higher molecular weights also make processing more difficult, so typically a balance has to be struck in optimizing the molecular weight in a given material.

Molecular weight distribution—in any given material, there is a large quantity of polymer molecules, and the length of these molecules (and their masses) is going to vary. The degree to which they vary is known as the molecular weight distribution. The goal of most polymerization processes is to have this distribution within a predictable and well-controlled range.

Melt cycle—a melt cycle is a transition of a given polymer from a solid state, to a molten state under heating (and usually shear), and then back into a solid state after cooling—one single time. Most thermoplastics can endure multiple melt cycles, however, each time a polymer goes through a melt cycle, the polymer chains tend to degrade slightly, and get broken down into shorter lengths. The net effect of a melt cycle is that the molecular weight distribution is skewed downward; most of the polymer chains are shorter, and the variation in length goes up. As a result, the viscosity of the molten resin changes, and the physical properties of the material changes as well. This phenomena is an important factor in recycling and reprocessing.

3.4 The Resin Industry

The resin industry is a unique subset of the plastics industry, and consists of the suppliers (producers), distributors, and resellers of thermoplastic materials. In the United States alone, over 100 billion pounds of plastic resin was produced in 2013.[3] That compares to about 4 billion pounds of steel,[4] and 140 billion pounds of concrete.[5] On the basis of actual volume (that is, the actual occupied space), plastic resin production might be the largest—since plastic resins usually have a much lower density than steel or concrete. On a financial basis, the numbers are not as easy to compare, but no matter how you measure it, the resin industry is a big business.

You may find it unusual to have a business overview in a technical book on material selection. I would argue that an understanding of the resin business is a fundamental component in thermoplastic material selection. When you go to a party, after you have said hello to the hosts and been introduced to some guests, the first question you usually ask is, *Where is the bar?* (followed by *Where is the food?* and *Where is the bathroom?*). At most parties, practically anyone can answer these questions.

In the world of thermoplastic material selection, the first question you usually ask is, *Where is the data?* The problem is, the answer to that question often depends on who you ask. In order to determine *who* you should ask, we need to understand where we are in the plastics food chain, where the other guests at the party reside in the food chain, and what vested

---

3. American Chemistry Council.
4. American Iron and Steel Institute.
5. United States Department of the Interior, U.S. Geological Survey.

interest they have in the answer. In a later chapter, we will also discuss who to ask for help when the first answer you get lands you in trouble.

3.4.1 Resin Production

As described earlier, over 100 billion pounds of plastic resin were produced in the United States in 2013. This represents what I like to describe as a Carl Sagan[6] number. It is a massive, mind-boggling quantity. In addition to some of the more obvious questions *(uhm, excuse me, but where did it all go?)*, there are other some other subtle questions that need to be asked (Figure 3.11).

Who is making this stuff? Why should I use it? How can I be sure I am using the right material, in an appropriate manner, in a way that ensures the health, safety, and welfare of the public? There are no simple answers to these questions—except perhaps for the first one, *Who is making this stuff?*

Thermoplastic resins are produced not only in the United States, but all over the world, on every continent except Antarctica. Some producers focus exclusively on resin polymerization. They buy feedstocks and catalysts and/or the chemical intermediates and then polymerize the final resin. Some resin producers are vertically integrated, and basically make everything from scratch. There are significant economies of scale in resin production, and, like in many industries, the highest volume producers typically have the lowest costs. As a result, resin production sites are often very large facilities that require a substantial amount of capital investment, and the industry is dominated by global conglomerates. There are very few resin producers making resin in someone's garage.

Rather than making every possible thermoplastic resin, most resin producers specialize. Some focus on polymerization technology, others on a chemical family, and others on specific market segments. Most resin producers are also involved in research and development. They employ chemists, chemical engineers, polymer scientists, physicists, and are constantly exploring new formulations, new additives, and new methods of polymerization. Some might argue that all the important thermoplastic materials have already been developed, a variation of the famous quote, *Everything that can be invented has been invented.*[7] Others might argue that there are breakthrough materials yet to be

---

6. Billions and Billions: Thoughts on Life and Death at the Brink of the Millennium.
7. C.H. Duell, Commissioner, U.S. patent office, 1899, (attributed).

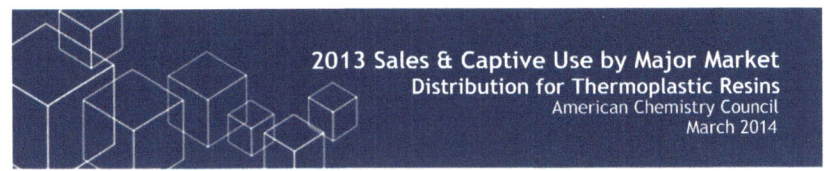

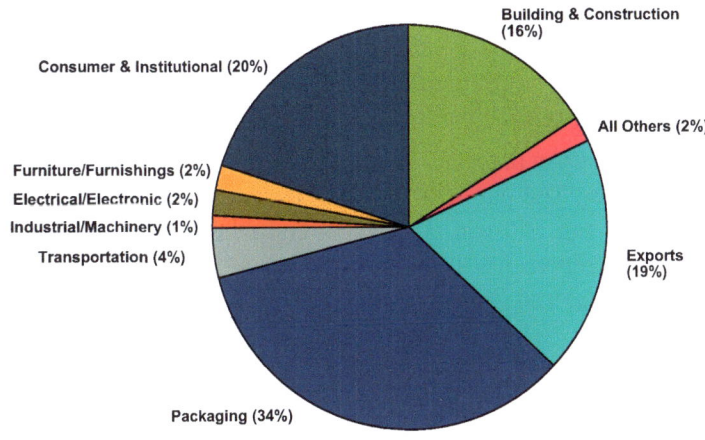

Figure 3.11 Plastics market use.

discovered. Regardless, the production of thermoplastics as raw materials is a very big party.

3.4.2 Resin Distribution

While the production of thermoplastic materials typically involves a massive amount of capital investment, the distribution of these materials follows the traditional paradigm for raw material transport. Distribution is typically done via road and rail freight, with the standard shipping units being a bag, bulk box, truckload, railcar, or boat load.

Bag—for whatever reason, a bag has become a universal standard of measurement for most bulk solids, regardless of whether it is a bag of flour, bag of fertilizer, bag of concrete, bag of coffee beans, or bag of plastic resin. The bags themselves are made from a variety of materials: paper, burlap, nonwoven synthetic fiber, etc. Regardless of the materials used, or the contents within the bag, a bag usually represents a quantity of material that a single adult can pick up and carry without assistance.

In the resin industry, a bag is usually made of a heavy paper, with a protective inner liner. The bag is filled with resin pellets, and then sealed, and labeled. A bag of resin typically weighs between 50 and 55 lbs (20–25 kg), and represents the smallest quantity for storage and transport of a given material (Figure 3.12).

Bulk box—a large box, typically of a standard size and designed to fit on a standard shipping pallet (a shipping pallet, also called a skid, is flat structured that provides support for the items being shipped, while also providing a means to lift and move and stack these items, usually via a fork lift, front loader, or other standard commercial shipping equipment. The shipping pallet can be made from a wide variety of materials, as can the actual box.).

A bulk box for plastic resins will normally hold 2200 lbs of resin (1000 kg, also called a metric ton). In the United States and Canada, a bulk box is sometimes referred to as a gaylord, as a reference to the Gaylord Container Corporation of St Louis, one of the first manufacturers of bulk box packaging. For many resins, a bulk box represents the minimum order quantity.

Note: a pallet can also be used to transport a number of bags of resin. Properly loaded, a pallet can easily transport 50 individual bags (Figure 3.13).

Truckload—a truckload is a generic term, used to describe the amount of plastic resin that can be transported in a single trailer by a commercial semitrailer truck. A truckload usually consists of a number of bulk boxes (or palletized bags), with a total weight of about 44,000 lbs (20,000 kg).

Railcar—a railcar (short for railroad car) is a generic term used to describe the transport of cargo on a rail transport system (a railroad).

3: Understanding Thermoplastics

Figure 3.12 Bags stacked on a wooden pallet. SlavoljubPantelic/Shutterstock.com.

Figure 3.13 Cardboard bulk box on a shipping pallet. elgusser/Shutterstock.com.

Figure 3.14 Covered hopper car, used for rail transport of bulk solids. ButskykhRoman/Shutterstock.com.

Plastic resin is usually transported in a specific kind of railcar known as a covered hopper car, which has a capacity of approximately 220,000 lbs (100,000 kg) (Figure 3.14).

Boat load—a colloquial term, often used to convey a really big quantity of something. In the resin industry, thermoplastic materials are often transported by boat, either via ocean freight, or river transport. Transport is usually done via the use of commercial shipping containers, which are rectangular and modular. Standard containers are 8 ft (2.4 m) wide and 8.5 ft (2.6 m) high, and come in lengths of 10, 20, 30, 40, and 45 ft (3, 6, 9, 12, and 13.7 m).

While a standard 20-ft container can typically hold about 10 pallets, either bulk boxes or palletized bags, it is often more economical to ship resin in a bulk manner, in which case a standard container can transport about 44,000 lbs (20,000 kg) of resin—the same as a truckload.

By the way, there is no standard measurement for a boat load of resin. It will depend on the shipping containers used, and the size of the boat.

As with all types of freight, the shipping costs for resin distribution are variable, and depend on the mode of transport, the distance involved, and quantity of material being shipped. By far the most efficient means of transport is to fill a railcar at the supplier's point of origin, and then transport it via rail (train), directly to the manufacturing site where it will be molded into finished parts. Of course, this is not always possible. While it works for high-volume applications, it does not allow for customization or modification, and is not practical for specialized resins or small production runs.

As a result, there is a complex infrastructure for resin distribution, involving not just the resin producers, but distributors and resellers, as well as recyclers and reprocessors. There are agents and brokers, warehouse managers, freight forwarders, customs brokers, and other logistics providers. There are also compounders (who buy in bulk and then modify and customize resins), stock shape providers (who buy resin in bulk and then fabricate standard shapes), and other intermediate product manufacturers (who also buy in bulk and then fabricate various products). This infrastructure is in many ways a classic example of supply and demand, and as a result the global supply chain has a predictable pricing structure, with the costs for distribution factored in.

From a material selection perspective, it is often important to consider *where* the resin is coming from, and *how* it is being delivered to the manufacturing site. These sometimes affect the material cost as much as *what* material you are using, and *who* made it.

3.4.3 Resin Grades

Like all raw materials, thermoplastic resins can also be classified according to grade. And like other raw materials, there are different methods of grading—some methods are based on international standards, some are based on visual comparison, some are based on a chain of ownership, and some are based on usage history (either actual or perceived). Some grading methods are even based on the type of processes the raw material has gone through prior to delivery to the molder.

Below is a list of terms describing some commonly used grading methods:

Prime—a resin from a certified producer that meets international standards for performance and consistency. Typically the seller of the resin (a distributor, a broker, or even the manufacturer themselves) will provide documentation of the certification process.

Generic prime—a resin that is certified to meet international standards for performance and consistency, but where the specific producer is not disclosed. Typically the seller of the resin will provide documentation of the certification process (Figure 3.15).

Food grade—a resin that is certified to meet international standards for safe contact with food. In addition to verifying the safety of the chemical composition of the resin itself, there must be no contamination of the resin during manufacture, and the material cannot degrade to become a source of contamination during use. Typically the seller of the resin will provide documentation of the certification process (Figure 3.16).

Figure 3.15 Certified generic prime. Aready/Shutterstock.com.

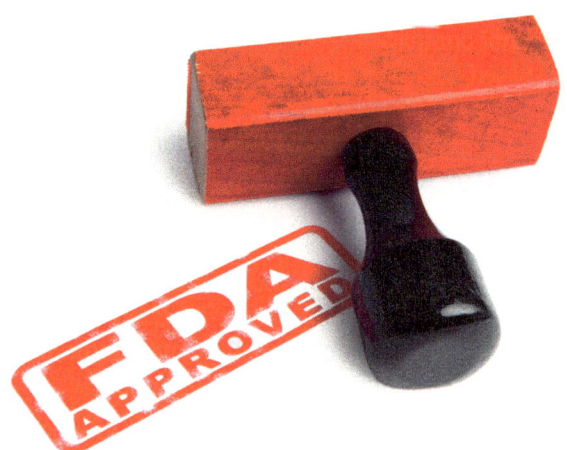

Figure 3.16 FDA approved. HurstPhoto/Shutterstock.com.

Medical grade—a resin that is certified to meet international standards for safe contact with the human body for medical applications. There are numerous subcategories and classes, depending on the nature of the contact (temporary vs long-term, intermittent vs constant, etc.). To ensure safety, these standards normally include adherence to strict protocols and require extensive documentation.

Figure 3.17 Off specification. Nata-Lia/Shutterstock.com.

Optical grade—a resin that is processed to provide consistent performance in the transmission of light, typically in the UV and visible light spectrums. Optical-grade resins have exceptionally high clarity, with no detectable discoloration or contamination.

Off-spec—(sometimes also called *wide-spec*) a resin that does not meet all of the published specifications. The deviation from the specification may be very small, and it may be in only one property, or the resin could miss a number of specifications by a wide margin (Figure 3.17).

Industrial—a subset of off-spec, usually implying that the deviation from specifications is small, and will not affect functional performance in most applications. For instance, the resin may have some discoloration, but maintains strength and consistency. In some cases, an industrial grade may come with documentation of what specifications are substandard, and by exactly how much; also often described as *commercial*.

Reprocessed—a resin that has been through one or more melt cycles. This may be due to compounding, melt blending, recycling, regrinding, etc.—or a combination of the above. Note: this term is widely used, and often loosely defined. Care should be taken to define and clarify the type of reprocessing that has occurred, and how many times the material has been reprocessed.

Recycled—a resin that has been through one or more production processes, and has achieved its end-of-intended use and been designated

Figure 3.18 Recycling symbol. nikolae/Shutterstock.com.

as waste. However, instead of being disposed of, this material is then sorted, cleaned, reprocessed and packaged for reuse (Figure 3.18).

Preconsumer waste—material waste that is a by-product of an industrial process.

Postconsumer waste—material waste from a finished consumer item that has reached the end of its intended use, and has been discarded by the user.

Regrind—a resin pellet that has been produced by chopping or grinding larger chunks of plastic that were produced during a previous molding process. These chunks could be remnants of the molding process (flash, runners, gate vestiges, etc.), nonconforming or rejected parts, left over material, etc. While technically regrinding is a type of recycling, the term regrind is commonly used to describe an in-the-factory process, and is rarely considered to be a recycling process.

Virgin—a resin that is in the original state as it was first packaged by the resin supplier. While it may be a prime or off-spec resin, it has not been reprocessed in any way. It is the industrial equivalent of a consumer buying something "brand new in the box." Many applications require the use of virgin resin (Figure 3.19).

The grading of thermoplastic materials is a poorly understood concept. In most industries, the word grade is used to quantify a specific level of

Figure 3.19 Virgin symbol. http://www.shutterstock.com/pic-128183369/stock-vector-grunge-rubber-stamp-with-the-olives-and-text-extra-virgin-olive-oil-vector-illustration.html?src=8TKEAj5fvlvZc121aSpdyg-1-10.

quality, consistency, and performance. Unfortunately, the plastics industry has a fast and loose relationship when it comes to describing grades of thermoplastics. This is NOT a statement about the quality, performance, or reliability of these materials. Indeed, the standards for measurement and quality of thermoplastic resins are among the highest of any materials industry. Rather, it is a statement about the *language* that is used within the plastics industry. One supplier will use the term to describe a group of materials as meeting a specific standard (i.e., a food *grade*), another supplier will use the term to describe how the materials are processed (i.e., a blow-molding *grade*), and yet another supplier will use the term to describe materials that have been modified for improved performance (i.e., a toughened *grade*). This blurring of the term grade can lead to confusion and misunderstanding.

Regardless of these issues, the selection of the proper grade of thermoplastic material can be just as important to the success of the application as the selection of the proper type of material.

3.4.4 Resin Modification

Thermoplastic, resins are frequently modified from their original as-polymerized condition. These modifications may be done for a variety of reasons, perhaps to aid in handling or processing the raw material, or to enhance one or more properties of the material in the manufactured part in its end-use environment.

Resin modifications are typically accomplished via the use of additives, which are introduced to the resin during a compounding or blending process. These additives may be in a liquid or powder form, or in the shape of small solid particles. Occasionally, another resin can even be used as an additive. The modifications typically affect the material on a mechanical level, such as a change in viscosity of the molten resin, or a change in the strength or stiffness of the material—as measured in the molded part. Some modifications can affect the electrical properties of the material, such as increased electrical conductivity. On rare occasions, these modifications can even affect the chemistry of the base resin.

There are numerous methods available for resin modification, including the following:

Compounding—a batch process where a given amount of material is blended with additives (or with another material) and then the entire batch is sent through a blending extruder and processed into new pellets. The advantage of compounding is the consistency and uniformity of the resin through the entire batch. The disadvantage is that the polymer has now gone through an additional melt cycle.

Melt blending—a continuous process, where a given polymer is run through a blending extruder and additives are introduced to the molten polymer under carefully controlled conditions. It has the same advantages and disadvantages as compounding, but is only suitable for large production runs.

Cube blending—another type of batch process, where a batch of pellets (aka cubes) of one resin are mixed with a batch of pellets of one or more other resins. The blending usually occurs at room temperature, in a large container, that is then shaken and/or tumbled. The advantage of this process is that is simple, inexpensive, and does not subject the resin to a melt cycle. The disadvantage is that it can be difficult to ensure a consistent mix throughout the entire batch (Figure 3.20).

Dry blending—similar to cube blending, dry blending is another type of batch process, where a batch of pellets (aka cubes) of one resin are mixed with other dry materials (powders, pellets, etc.). The blending usually occurs at room temperature, in a large container, that is then shaken and/or tumbled. The advantage of this process is that is simple,

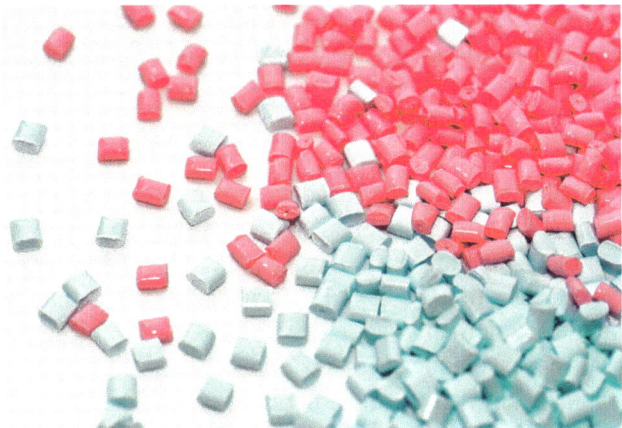

Figure 3.20 Resin cubes of two different colors, about to be cube blended. XXLPhoto/Shutterstock.com.

inexpensive, and does not subject the resin to a melt cycle. The disadvantage is that it can be difficult to ensure a consistent mix throughout the entire batch.

Resin modifications can occur in the original production facility where the resin was first polymerized, but they can also be modified at a later date at a different facility. Some modifications can even be performed on the factory floor, just prior to, or during, the processing of the resin into production parts. However, the most common method of modification is via compounding. There is an entire industry built around this process. Compounders typically buy generic prime resin from a distributor, and then modify it with various additives—usually to create a specific amount of material for a specific customer order.

Additives typically fall into one of four main categories: processing aids, fillers, reinforcements, and performance modifiers.

3.4.5 Processing Aids

The most common use of additives are to aid in specific areas of processing. Many of these additives are hidden from sight, as it were, as they are designed to affect the viscous properties of the molten resin, to help it flow better, or to be more homogeneous, or to cool and solidify (and shrink) in a more consistent manner. Some additives are intended to preserve the equipment used in plastic processing, to prevent corrosion of the metals, or enhance the release of the solidified plastic from the mold surfaces. Some processing aids are specific to a certain type of plastics processing, while others are suitable for a variety of processes.

3.4.6 Fillers

Fillers are additives that are used to replace some of the volume of the resin with a filler material. The filler material is typically a low-cost generic item (such as talc), but it may also be a specialty item. Fillers are typically added in a fiber or powder form, and may include talc, mica, glass beads, glass microspheres, glass fibers, etc. The most common use of fillers is as a low-cost means of making parts stiffer.

Fillers can also be used to adjust the density of the final part. The addition of small, hollow glass spheres can be used to reduce the density of an injection molded part by up to 20%—without the disadvantages of structural foam molding. Conversely, the density of a molded part can also be increased by adding small particles of a high-density material, such as tungsten.

The materials used as reinforcements and fillers are often similar, yet there is a significant difference. In a reinforced resin, there are chemical bonds between the additive and the polymer chains. The strength of these chemical bonds greatly increases the strength and stiffness of the material—they are reinforced. There may even be proprietary technology used to create superior coupling, to enhance the performance even further. In a filled resin, there are no chemical bonds between the filler and the polymer chains (or they are so weak that they can be ignored). While a filled material is often stiffer than an unfilled resin (in a similar manner that chicken broth thickens into gravy with the addition of flour), rarely is it stronger. In fact, in pure tension, it is often weaker.

3.4.7 Reinforcements

Reinforcements are additives that are used to increase the structural performance of the base resin. Typically the reinforcing materials are specialty items, and while they may be in a powder or bead form, the most common shape is a fiber. Common fiber-reinforcing agents include glass fiber, carbon fiber, and aramid fiber.

Most of us are familiar with fiberglass or carbon–fiber composites. In each of these, long fiber strands are woven to create a cloth. This cloth is then draped over a form (a surfboard, a mold, etc.), and then saturated with a thermoset epoxy resin. This resin then cures, and in the process bonds to the fibers in the cloth. The net result is a product that is stiff and strong, yet also very light. In addition, carbon–fiber composites are often considered cool and sexy.

A reinforced resin is the thermoplastic equivalent of thermoset-based composite. One might even call such materials *thermoplastic composites*. One limitation of reinforced resins is in the length of the fiber that can be used. In a thermoset application, one can preform the cloth into a desired shape before introducing the liquid resin. In a thermoplastic application, the fibers must somehow be mixed into the molten resin. This is usually done in a compounding process, and the resulting reinforced resin is then cut into pellet form, so the overall length of the reinforcing fiber is typically limited to the length of the pellet itself. Even with this limitation reinforced thermoplastics can offer some extraordinary physical properties (Figure 3.21).

Compared to its base resin, a reinforced thermoplastic will be stiffer, stronger, and often tougher as well. This is due not only to the properties of the reinforcing agent, but to the strength of the chemical bonds between the polymer and the reinforcement. The chemistry of these bonds is a science all to its own, indeed it is often a source of competitive advantage between various suppliers. Supplier A may have developed a superior bonding technology, so that even though supplier A and B are using the same base resin and reinforcing agents, the material from supplier A will have better end-use performance.

3.4.8 Performance Modifiers

In addition to structural reinforcement, additives can also be used to enhance other aspects of material behavior. Some of these enhancements

Figure 3.21 Glass fiber "chopped strands" which are commonly used for reinforcement of thermoplastics. *Photo by Holger Casselmann, Wiki commons.*

are minor, some can be quite substantial. Stabilizers can be added to improve resistance to UV degradation, or exposure to high heat. Tougheners can be added to improve impact resistance. Lubricants can be added to reduce friction. Pigments and dyes can be added to enhance the visual appeal.

The ease of modification is one of the reasons that thermoplastics are so widely used. Regardless of the additives used, it is important to remember that they can sometimes have undesirable consequences. They may affect mold shrinkage, act as unintended nucleating agents, or have a trickle down effect in other areas of performance.

3.4.9 Resin Versions

Once a resin has been modified, it is usually given specific product code (which is often unique to the supplier or compounder), as well as a descriptive common name. While many resin suppliers will describe these modified resins as being a specific *grade*—I believe this is an incorrect use of the term. Unless the modifications are done in compliance with specific international standards, it should not be described as a grade of resin.

The term I prefer to use is *version*. A version is a variation of something, where that something has been modified. Let us say that the thing you are modifying is a car engine, and you modify a certain way, and then your next door neighbor asks you what the modification is. So you tell him, "It's a chop top, flat block, slant 6 that's been modified for drag racing but it's still legal for street use." So what grade of engine is it? I would call it a modified version.

Below is a list of terms describing some common methods of resin modifications:

Neat—a resin that has not been modified in any way from its original as-polymerized condition—either via reprocessing, compounding, blending, or any other secondary process.

Natural—a resin where the color has not been modified from its original as-polymerized condition. The natural color of a resin depends on its chemistry. Some resins are naturally clear; some are naturally white, while others are off-white, brown, or black.

Colored—a resin where the natural color has been modified through the use of coloring additives such as dyes, pigments, etc.

Special effects—a resin where the appearance has been modified through the use of different additives for special visual effects, such as metallic flakes for sparkle, or insoluble dyes for a swirled look.

General purpose (also called *multi-purpose*)—a resin that has been modified to serve as many uses as possible. It is also used to describe a resin that is suitable for most standard uses of that material.

Injection molding—a version that has been modified specifically for the injection molding process.

Extrusion—a version that has been modified specifically for use in an extrusion, or extrusion blow molding process.

Lubricated—a version with additives (such as silicone or molybdenum disulfide) that will provide increased lubricity in the fabricated parts. This can be helpful for parts such as gears and bearings.

Wear resistant—similar to lubricated, but with additives that provide greater wear resistance in the fabricated parts.

Nucleated—a version with additives intended to promote uniform crystallization in the molded parts (typically this only applies to semi-crystalline resins).

Toughened—a version with additives designed to increase the fracture toughness of the resin. There may be levels of toughness, from standard toughened to highly toughened to materials described as "super tough." Also known as *impact modified*.

Filled—a resin with a filler added.

Reinforced—a resin where the additive is bonded to the base polymer, and enhances the strength and stiffness.

Flame retarded—a resin with additives specifically chosen to reduce or eliminate burning, and/or melting and dripping when exposed to open flame.

Weatherable—(also called UV stabilized) a resin with additives to prevent discoloration and/or degradation from exposure to sunlight, primarily in the UV region.

Heat-stabilized—all organize polymers are subject to oxidation, and this oxidation increases with heat. Heat stabilizers reduce the rate of oxidation, allowing the resin to better withstand exposure to heat, either during processing, or in the end-use environment. Also know as "high heat" or "high temperature" versions.

Plasticized—a version with special additives called plasticizers. These additives improve the flexibility of the molded parts, especially at low temperatures. The "new car smell" in an automobile is typically the result of plasticized resins.

One of the challenges with modified resins is that they are often difficult to compare. One supplier may modify a resin one way while another supplier modifies it another way, but they both call their new modified resin by the same name. This is a classic example of comparing apples to oranges. To reiterate my earlier comments about grades, a grade should imply that a certain material complies with an external standard, and a resin from supplier A that meets that standard is equivalent to a resin from supplier B that meets that standard.

3.4.10 Alloys and Blends

One of the most interesting aspects of thermoplastic materials lies in their ability to be modified. While this is often done through the use of additives, and it can also be done through the use of another thermoplastic material. It is quite common to mix one or more thermoplastic materials, either in the lab, during the resin production process, and sometimes even on the factory floor. The mixing can be done by combining solid pellets (commonly known as cube blending), or by mixing materials while they are in a molten state (either via melt blending or compounding).

As a result of this kind of mixing, one of several things can happen. Sometimes, the materials might just not mix together very well. In this case, they probably are not chemically compatible. If you tried to mold parts from this mix, you would most likely end up with a swirled looking part that simply separates at the boundaries between the materials.

On most occasions, the two materials will mix together and make a blend. A polymer blend is the plastic equivalent of a salad dressing made of oil and vinegar. You take two different materials, mix them together, and then pour them over your salad. If you start with good ingredients, and then mix them well, more often than not you will have a nice tasting salad. A polymer blend is no different. If you start with good ingredients, and then mix them properly, more often than not you will end up with a delicious new material (Figure 3.22).

There are lots of polymer blends available, and more are being developed every day. Some of the most successful commercial applications of thermoplastics have occurred as the result of blends. The world's first plastic BMX bike wheel was made from a blend of 33% glass reinforced nylon 6/6 mixed with a sample of unreinforced but highly toughened nylon 6/6. The resulting blend had just the right amount of strength, stiffness, and toughness.

Figure 3.22 Oil & vinegar salad dressing—one of the world's greatest blends. Quayside/Shutterstock.com.

Another result of mixing two materials—and in many ways this is the preferred result—is that the two materials will have chemical affinity for one another, and will combine to form a new material with properties greater than the sum of the parts. This is a thermoplastic alloy. In a similar manner to how alloys of iron and carbon (steel) or of copper and tin (bronze) enabled new products and technologies in the Age of Metals, thermoplastic alloys opened an entire spectrum of new possibilities in the Age of Plastics. The combination of material variations, additives, possible alloys, and potential new applications is virtually unlimited. The next breakthrough thermoplastic material may very well be right in front of you.

3.4.11 So Where Is the Data?

Let us revisit the question, *Where is the data?* and more specifically, *Who do I ask?* First and foremost, the answer depends on where your application lies. Most applications can be roughly categorized into one of four categories (Figure 3.23).

Figure 3.23 One of the greatest challenges in selecting thermoplastic materials is in finding the design data.

The classics—these are classic applications of standard materials. Most have been around for decades. The material properties meet industry standards—and in many cases the standards were created based on the use of the materials in that industry. The materials have global availability, there is minimal differentiation between suppliers, and there is established market pricing based on supply and demand. In this scenario, the data is readily available, and you can ask anyone—the resin producers, the distributors, even an online database.

The classics, tweaked—these are also classic applications of standard materials, but slightly tweaked. Perhaps you are making sporting goods equipment, and were using a glass-reinforced nylon, but need a little more toughness. Or perhaps want to make a cell phone housing in fluorescent pink using an ABS material, and no one has done that before. But ABS has been in color for decades, including some other fluorescent colors, and can be done.

In this scenario, the dance for data can be a bit delicate. In some cases, the data is simply unavailable. So it does not matter who you ask, it just does not exist. In other cases, there may be some data available, but it is incomplete. Accessing this kind of data can be challenging—there may be legal issues, confidentiality issues, etc. So while it does not hurt to ask, do not be surprised if there is no response. In other cases, which happens more often than you might think, the tweak in question represents something unique and proprietary, and the producer or compounder has a vested interest in NOT sharing that data with you.

It allows them to provide a material with a competitive advantage without disclosing the actual results of what they have actually done. Sometimes they may share that data with you in a confidential manner (the industrial equivalent of *Don't Ask, Don't Tell*), but there usually has to be a very strong relationship before that occurs.

New technology—this category represents the latest and greatest of what is happening in the world of thermoplastics, including new applications, new materials, and new derivative materials. In this scenario suppliers are normally eager to share data, as it can help develop an entire new market for their materials. In the 1980s, it was the use of engineering plastics in automobiles. In the 1990s, it was thermoplastic encapsulation of electronics. In 2000, it was the use of thermoplastics in the wireless industry. The interesting thing is that today's new technology becomes tomorrow's classic. So in this area, ask anyone you come across. If the data exists, more often than not, they will readily share it with you. (In return, you should allow that supplier to promote your use of that material in their marketing efforts. It is good plastics karma.)

Secondary markets—this category represents everything else. It involves the use of off-spec materials, industrial grades, resellers, and recyclers and everything else. The challenge is not just who to ask for the data, but if there is any data at all.

In all of these scenarios, the infrastructure in the resin supply chain is often your first line of inquiry. Most of the major resin suppliers have customer support centers, with toll free numbers, as well as extensive Web sites with links and search engines. They also have sales and development specialists (many who may focus on an industry or a specific technology), and technical support staff. The distributors and compounders have a similar structure, and help is never more than a phone call away.

Overlaid on all this of course, is the unspoken question, *Who is asking?*

If you are a senior staff engineer for a major international consumer electronics manufacturer and are making an inquiry about property data for various materials regarding transmission and absorption effects in the RF spectrum, you will probably get an answer fairly quickly. It will most likely involve a technical report with some charts and tables, and should have some data that is relevant to the specifics of your application. This report might be a reprint of a technical paper that was presented at an international conference, or it might be white paper labeled "Proprietary and Confidential."

Figure 3.24 VIP section. heromen30/Shutterstock.com.

However, if you are an independent consultant working for a manufacturer of small peripherals and ask the same question, you probably would not get much of a response. Do not take it personally. It is a big party after all, and not everybody gets a VIP ticket (Figure 3.24).

3.5 Thermoplastic Classification Methods

In much the same way that humans have always made things, humans have always tried to classify things. We sort and categorize, compare and contrast, arrange and rearrange. It is one of the ways we try to understand the world around us. It is human nature.

This drive to classify things is evident in every aspect of our lives. We have systems and methods of classification for everything. In the field of biology, there is a comprehensive classification system to sort and organize life itself, and every living thing on our planet has a unique place in that tree. This system has been in development for centuries. It has its origins in the work of Carl Linnaeus, and is still being refined today (Figure 3.25).

Unfortunately, the world of thermoplastics does not have a tree of life, yet. Yes, there are chemical families and types and grades and versions, and there are categories and divisions and subdivisions. One can take a specific material, make some assessments, and place it somewhere on a giant grid, and then take other materials and do the same thing. And while many of these groupings will make sense, there comes a point where the current classification system breaks down.

Just like the discovery of the platypus wreaked havoc with the biological classification system of the time (it was initially named *Ornithorhynchus paradoxus*), the development of new plastic materials continues to wreak

3: Understanding Thermoplastics

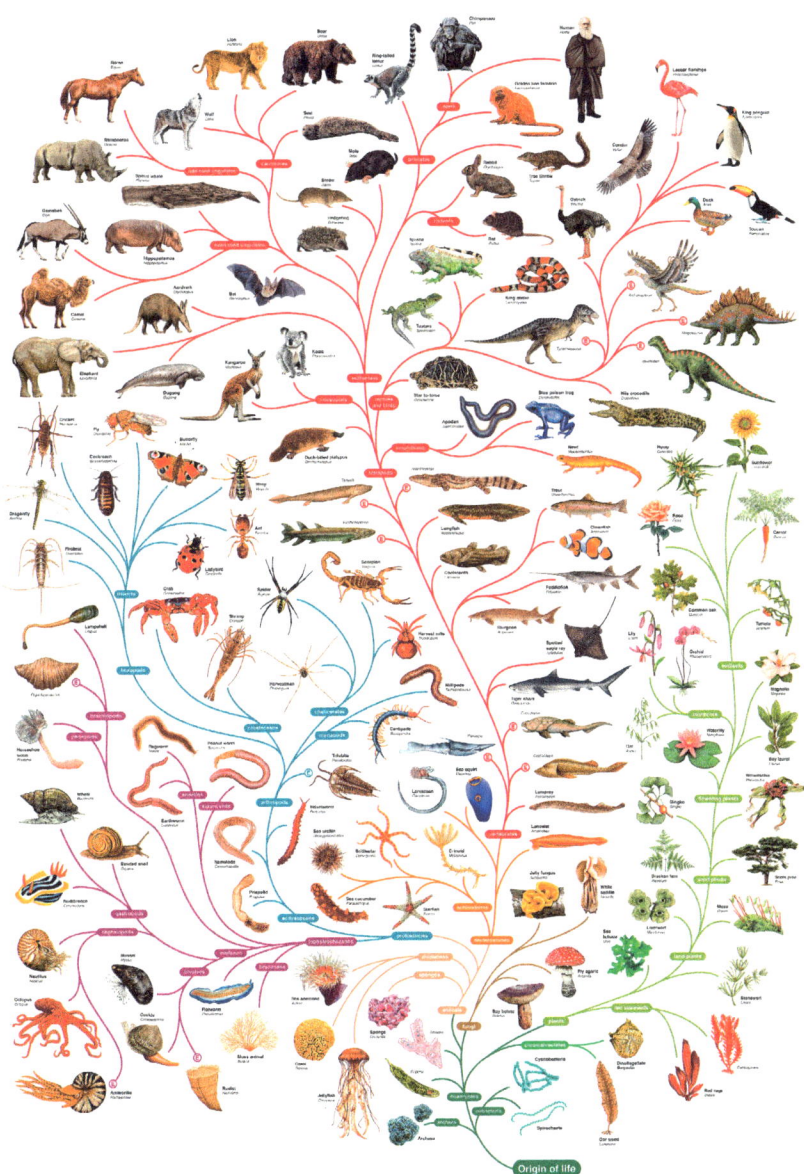

Figure 3.25 The tree of life.

havoc with "standard" plastic classification methods. How can a TPU be both a thermoset, and a thermoplastic? How can a cross-linked material be a thermoplastic? (I thought you said cross-linking only happened in thermoset materials.) How come this polyamide is semicrystalline, and this polyamide is amorphous? *And how am I supposed to remember all this?*

Once again, there are no simple answers. Plastics are unique materials, and the categories used to describe and compare and contrast them are not always black and white. There are some high-level distinctions such as thermoplastics versus thermosets (described earlier), and amorphous versus semicrystalline. These distinctions are useful, and important, but by no means absolute.

3.5.1 Amorphous versus Semicrystalline

As discussed earlier, amorphous thermoplastics have polymer chains arranged in a random, haphazard state. They have no distinct melting temperature, but instead soften when heated above their glass transition temperature. As a class, their properties differ from the properties of semicrystalline thermoplastics in some general ways (lower strength, less chemical resistance, etc.). These differences are well documented and frequently discussed, but I believe these differences are over exaggerated.

The real difference lies in processing. Since they have no distinct melting temperature, there are some processes, such as rotational molding, where you cannot use an amorphous thermoplastic. In the injection molding process, amorphous materials are sometimes described as being "easier" to mold. Since they do not undergo a phase change (i.e., from a liquid to a crystalline state), the mold shrinkage for an amorphous plastic is usually less than that of a semicrystalline material, and is primarily dependent on thermal contractions due to changes in temperature and pressure. This *may* result in a molded part that is flatter and straighter than a part molded from a semicrystalline material. However, the injection pressure required to fill a mold cavity with an amorphous material is almost always higher than what is required with a semicrystalline material. As a result, a part made with an amorphous "easy-to-mold" resin may have higher molded in stress in the final part, and will fail sooner.

In a semicrystalline thermoplastic the polymer chains are arranged in a regular, repeating (crystalline) pattern. They have a distinct melting temperature, and also experience a change of properties when heated above their glass transition temperature. As a class, their properties differ from the properties of amorphous thermoplastics in some general ways (higher strength, better chemical resistance, etc.), but again I think these differences are exaggerated.

Again, the real difference lies in processing. Since they have a distinct melting point, most semicrystalline materials are well suited to injection molding. Upon the application of heat (and some shear), they will melt into a liquefied state, and then with the application of pressure they easily flow and fill the mold cavity (usually with less-injection pressure that what is needed for an amorphous thermoplastic), and then cool and rapidly solidify. Since they will undergo a phase change (from a liquid to a crystalline state), the mold shrinkage will be higher—the shrinkage will include not only the thermal contractions due to changes in temperature and pressure, but also the change in volume due to the phase change. (The phase change usually accounts for the majority of the total shrinkage.)

Going back to the concept of an "easy-to-mold" material, most conversations about this have to do with controlling the shape of the final part, its overall size, dimensions, flatness, straightness, etc. There are those who claim that this is easier to do with a low-shrinkage material. However, it can also be done with a high-shrinkage material—as long as the shrinkage is uniform. In a semicrystalline material, this means that the amount of crystallization, and the rate of crystallization, needs to be carefully controlled.

What is interesting is that there is this big distinction between amorphous and semicrystalline materials, but very rarely is there a distinction among the various types of semicrystalline materials, specifically in regards to their amount of crystallinity (ranging from 0% to 100%). If crystallinity is that big of a deal, why do not we categorize materials based on that?

There are some processes that semicrystalline materials are not suited for. For example, it can be very difficult to thermoform nylon or acetal. It is possible, but it is much more difficult than thermoforming ABS.

3.5.2 Chemical Family

Another common method of classification is to describe the chemical family of the resin, primarily in reference to the chemical family of the base monomer. This is often a primary means of classification within the plastics industry, since the chemistry is interrelated to the feedstocks and catalysts, the chemical intermediates, and the method(s) of polymerization. However, it also plays an important role in material selection, as the chemistry of the polymer plays an important role in the properties of the material.

3.5.3 Cost versus Performance

Another useful method of classification is to organize materials based on an evaluation of price versus performance. Materials at the lower end of the cost spectrum are often referred to as commodity materials, those in the middle are often referred to as engineering plastics, and those at the high end are often referred to as specialty plastics. This method of classification is commonly used within the industry, and is discussed in greater detail in the next chapter.

3.5.4 Elasticity

Another means of classification is to organize materials based on their flexibility. We often have specific uses for materials that are stiff and rigid, and other uses for materials that are flexible and elastic. While this kind of classification is often helpful on a general level, in reality it often becomes cumbersome, since many materials can be formulated in a variety of stiffnesses. It is mostly used to describe a general category of highly flexible materials, known as thermoplastics elastomers. These will be discussed in greater detail in the next chapter.

3.6 A Final Word about Property Data

Property data for the characterization of a given thermoplastic material is generated under highly controlled conditions. First of all, the test specimens have a defined shape. They are made from a prime grade of material, usually with no additives of any kind. The processing conditions are highly controlled, and if any secondary operations are employed (CNC machining, laser or water jet cutting, etc.), the specimens are thoroughly annealed to remove any internal stresses. These specimens are then subjected to the prescribed test procedures, and the resulting data is published as "typical property data." Perhaps somewhere in the footnotes there should be a disclaimer, *This is as good as it gets. Your actual mileage may vary.*

3.7 The Amazing World of Thermoplastics

So here we have these new materials called thermoplastics. They have properties that are different in every orientation, different in every direction (and often different at different places in a given direction), different at every temperature, and different for every rate of loading. The properties will also vary depending on what processing technology is used.

Even though all of these properties are predictable and repeatable, since the materials have nonlinear behavior in the elastic region of the stress–strain curve, most of the classical engineering equations for mechanical behavior can only provide an approximate result. So, even if you could accurately determine the correct loading conditions, find the correct property data *under those conditions* for the material being considered, and then use a classic equation to try to predict actual, end-use performance, you will only have a *should-be-close-to-the-real-world* answer, with no real way of knowing the margin of error. To make matters worse, you also do not know if the properties of the material in the actual parts *as-they-are-fabricated* are anywhere near the properties of the material in the test specimen that was used in the lab to generate the property data in the first place.

Welcome to the world of thermoplastics.

We'll be here all month, Sunday through Saturday, with have two shows daily, at 6 and 11. Enjoy the show. And please remember to tip your wait staff.

Further Reading

[1] Eugene Avallone, Theodore Baumeister, Ali Sadegh, Mark's Standard Handbook for Mechanical Engineers, eleventh ed., McGraw-Hill Professional, 2006.
[2] M. Berins, Plastics Engineering Handbook of the Society of the Plastics Industry, Springer Publications, 1994.
[3] Stephen Engelsmann, Valerie Spalding, Stefan Peters, Plastics, Birkhäuser Architecture, 2010.
[4] Christopher C. Ibeh, Thermoplastic Materials: Properties, Manufacturing Methods, and Applications, CRC Press, 2011.
[5] Tim Osswald, Georg Menges, Materials Science of Polymers for Engineers 3E, Hanser, 2012.
[6] A.Brent Strong, Plastics: Materials and Processing, Prentice Hall, 2005.
[7] Stephen Timoshenko, Strength of Materials, Part I, Elementary Theory and Problems, D. Van Nostrand Company, 1955. first ed. 1930, second ed. 1940, third ed. 1955.
[8] Stephen Timoshenko, Strength of Materials, Part II, Advanced Theory and Problems, D. Van Nostrand Company, 1956. first ed. 1930, second ed. 1941, third ed. 1956.

4 An Overview of Thermoplastic Materials

If you look at the world of thermoplastics from a user perspective—that is, as someone who is going to select and use a single, specific material—you are faced with a daunting task. Not only are there hundreds of different types of materials, when you add in all of the different grades and versions you will have a list that numbers in the tens of thousands. How are you going to sort through these materials and find the one you need? In some ways, it is even worse than finding a mate. Sure you might use some online tools and run some sorting protocols to determine what materials might meet certain performance criteria, but there are easier ways.

One of the easiest ways to select materials is to begin by organizing them based on their relative cost. Like most things in life, materials that are lower in cost are typically used in greater quantity, while more expensive ones are typically used less. You can also organize materials based on their performance level. The performance can be based on temperature requirements, or structural needs, or chemical compatibility. This kind of sorting is not data dependent, but is instead based on a general sense of *good, better, best*. If you undertake these tasks in unison, an interesting pattern will emerge. Materials that are low in cost typically offer good performance, materials that are moderate in price offer better performance, and the most expensive materials offer the best performance. Once again, like most things in life, you get what you pay for (Figure 4.1).

4.1 Key Thermoplastic Materials

While there are literally tens of thousands of different thermoplastic materials, the reality is that there are really just of handful of basic material types. If you group these material types by their chemical composition, their chemical family if you will, you can begin to understand the family as a whole, and then begin to explore the subtle differences between various members of the family… the brothers and sisters and nephews and nieces and cousins. Fortunately, most materials within a given chemical family are usually in the same price-to-performance category.

In this chapter, we will discuss some key thermoplastic families, grouped into three price-to-performance categories: commodity plastics,

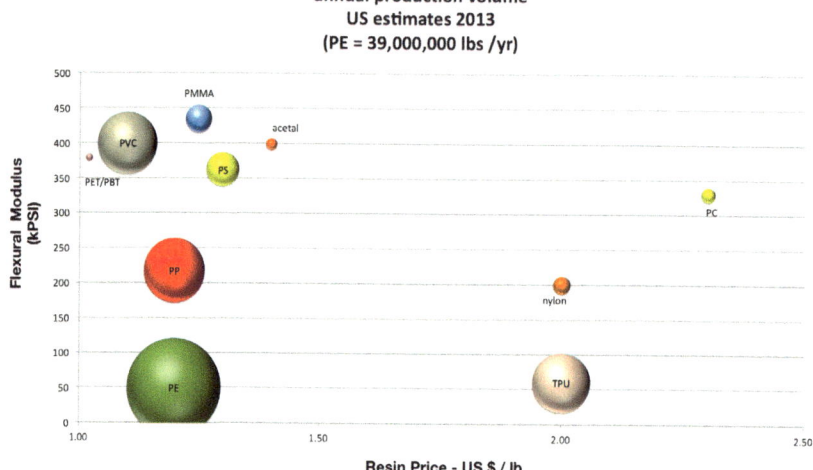

Figure 4.1 Price–performance–volume chart for commodity and engineering materials.

engineering plastics, and specialty plastics. Then, we will discuss the key families in a special category called thermoplastic elastomers (TPE). This discussion is intended to provide you with a general description of each material family, along with guidance on where—and why—the materials in that family are often used (Figure 4.2).

4.1.1 Commodity Plastics

The term commodity plastics is used to describe a category of thermoplastic materials that are widely used and readily available around the world—hence the term commodity. While they can be used for structural purposes, they are typically used in cost-sensitive, high-volume applications such as packaging, clothing, and personal items intended for short-term use. Key commodity plastics include the following.

Acrylic

A family of plastics technically known as polyacrylates. Derived from acrylic (or methacrylic) acid, the family includes a variety of materials, including thermoplastic and thermoset resins, textile fibers, paints, and adhesives. Polymethyl methacrylate (PMMA) is one of the most important. Another interesting member of this family is cyanoacrylate, better known as super glue.

PMMA was first introduced by Rohm and Haas under the trade name Plexiglass®. With a moderate level of engineering performance and

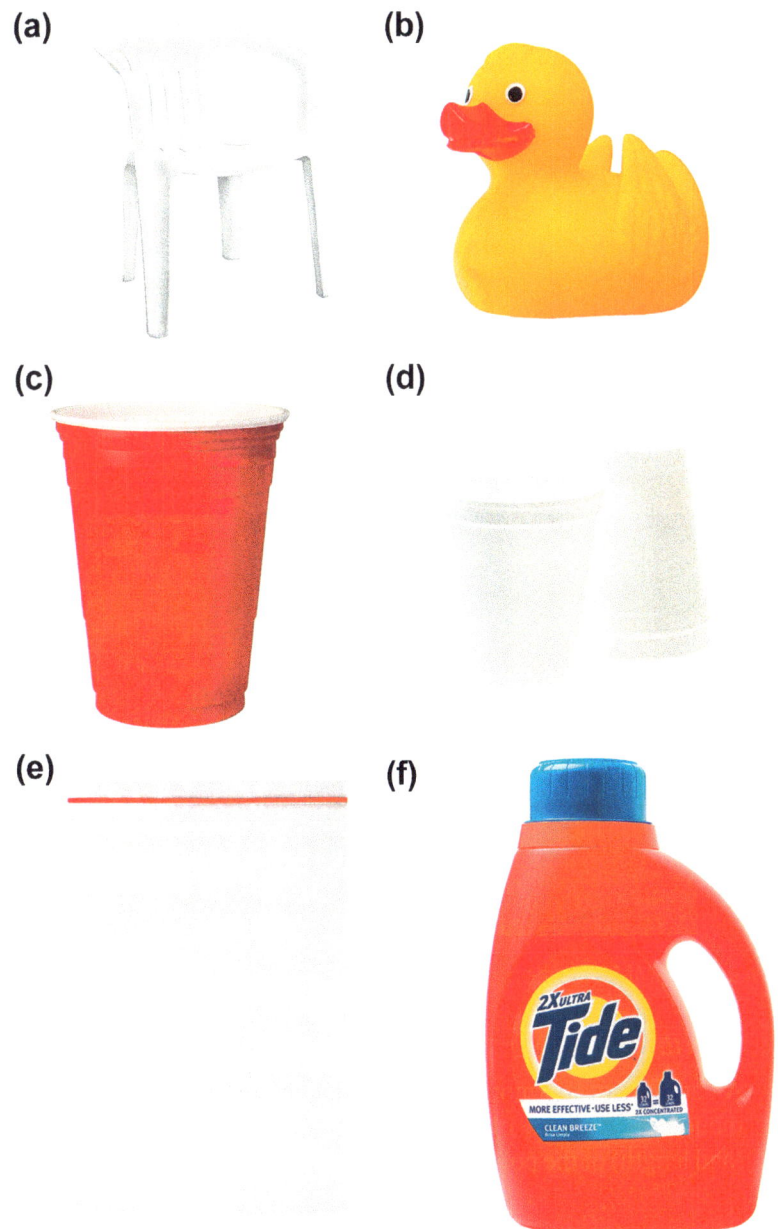

Figure 4.2 Some examples of products made from commodity plastics. (a) Polypropylene chair (Coprid/Shutterstock.com), (b) rubber ducky (Pabkov/Shutterstock.com), (c) Solo® plastic cup (DannySmythe/Shutterstock.com), (d) Styrofoam cups (design56/shitterstock.com), (e) Zip-Loc bags (koosen/Shutterstock.com), and (f) Tide detergent bottle.

decent mechanical properties, it is a relatively low-cost material that is easy to process. Among its capabilities, it can be cast, extruded, injection molded, laser cut, machined, and polished.

The most important thing to remember about PMMA: it has exceptionally high transparency, and perhaps the highest optical clarity of any thermoplastic. It is used as a replacement for glass in a wide variety of industries, including residential and commercial aquariums; cell phone screens and camera lenses; automotive lighting; commercial airline windows; and contact lenses. In consumer electronics, PMMA often is used in sheet form. It is almost perfectly flat and optically pure. A hard coat may be added to improve scratch resistance. The sheet then can be cut to size via a water jet, laser, or CNC machining (Figure 4.3).

Figure 4.3 PMMA is often used in aquariums. JaySi/Shutterstock.com.

Polyethylene (PE)

A category of thermoplastic materials that are polymers of the base monomer ethylene. PE polymers usually are classified by the type (and length) of the polymer chains. These types include the following:

- Low-density polyethylene (LDPE);
- Linear low-density polyethylene (LLDPE);
- Very low-density polyethylene (VLDPE);
- Medium-density polyethylene (MDPE);
- Cross-linked polyethylene (XLPE);

- High-density polyethylene (HDPE);
- High-density cross-linked polyethylene (HDXLPE);
- High molecular weight polyethylene (HMWPE);
- Ultrahigh molecular weight polyethylene (UHMWPE); and
- Ultralow molecular weight polyethylene (ULMWPE).

Polyethylenes are members of the polyolefin family. Polyolefins are polymers of olefin monomers, which are more commonly known as alkenes. As a class, polyethylenes are the most widely used thermoplastics in the world, for good reason. They have good mechanical properties, wonderful chemical resistance (they are semicrystalline), are very low in cost, and are easy to process. Among other processes, polyethylenes can be blow molded, extruded, injection molded, melt-cast, heat-staked, rotationally molded, thermo-formed, melt-swaged, thermally fused, and ultrasonically welded. They are used in consumer goods, in industrial and medical applications, and in packaging—in both rigid and film form. PE film is used for protective tarps, for grocery bags, even for bubble wrap packaging (Figure 4.4).

For design engineers, one of PE's more interesting traits is its surface characteristics. While relatively soft materials, they have excellent lubricity and good wear resistance. This makes them good choices for gears, cams, and mechanisms where the forces are relatively low. Polyethylenes are chemically similar to alkanes, also known as paraffins. To the layman, the term "paraffin" means wax. Rub a finger

Figure 4.4 Bubble wrap packaging made from polyethylene film. PavelHlystov/Shutterstock.com.

on a part made of PE and you will sense that is has a waxy feel to it. Wax on, wax off—go polyethylene!

Polypropylene (PP)

Polypropylene is another member of the polyolefin family. It is a semicrystalline thermoplastic material whose blend of properties leads to a wide variety of applications, ranging from commodity use to specialty engineering. While it offers reasonably good mechanical properties, it also offers exceptional chemical resistance at a relatively low cost.

The two main types of PP are homopolymer and copolymer, with homopolymer PP being the most widely used type. In general, it is stiffer and stronger than the copolymer type. Copolymer PP generally is softer than homopolymer, but is tougher and more durable, with better low-temperature toughness. Copolymer is usually divided into subgroups: i.e., block copolymer, random copolymer, and impact-modified copolymer. All of these types of PP are readily available from a wide variety of suppliers and distributors.

- *Block copolymer* consists of two or more homopolymers bonded together in a linear pattern (the term "block" refers to the distinct units that are bonded together). The properties depend on the pattern and ratio of the blocks.
- *Random copolymer* consists of two or more homopolymers bonded together in a random pattern. In PP, a random copolymer offers a balance of toughness and transparency.
- *Impact-modified copolymer* typically is a block copolymer with a rubbery component as one of the components of the matrix. In PP, this rubbery component typically is ethylene propylene diene monomer (EPDM), a synthetic rubber.

Polypropylene materials are typically stiffer and stronger than PE, but not as flexible. They are often used as structural materials in low-cost consumer applications, such as paint buckets, large pails, trash bins, and even recycling containers (Figure 4.5).

Polystyrene (PS)

Polystyrene is another widely used plastic with billions of pounds produced annually. Brittle, clear, and hard, it is made from the liquid styrene monomer, and is inexpensive on a per-unit weight basis. PS normally is extruded, injection molded, or vacuum formed. Chemically, it is considered a liquid hydrocarbon because it is composed

Figure 4.5 Recycling bin molded from polypropylene. SeanD/Shutterstock.com.

exclusively of carbon and hydrogen. It is in a solid state at room temperature and in a liquid state at about 100 °C. When burned, what remains only are black carbon particles, i.e., soot. Only water vapor and carbon dioxide remain when PS is oxidized completely.

In use, it either can be foamed or rigid, colorless (its natural state) or colored. Expanded PS foam is better known as Stryofoam™. Widely used for insulation purposes, its uses include containers, packing peanuts, tumblers, disposable cutlery, and trays. Further, because of its buoyancy, it is used for US Coast Guard-approved life rafts, as well as surfboards. A good heat insulator, it is commonly used as foam for take-out, clam-shell containers at restaurants, or even as a layer under pavement to prevent freezing and thawing of the soil, thus retarding surface cracking. Given its light weight, it also is used as a backing for photos or to construct architectural models. It is used to make model cars and airplanes, drinking cups, and appliance knobs (Figure 4.6).

Polyvinyl chloride (PVC)

Also known as vinyl, PVC is an amorphous thermoplastic formed by the polymerization of the monomer vinyl chloride. It is low in cost, easy to process, and broadly used. Its uses range from drain and potable piping to phonograph records (remember those?!?). Additional uses include toys, credit cards, rain gutters, flooring, binders and pens, jacketing on electrical wiring, and garden/patio furniture. In short, it is everywhere (Figure 4.7).

PVC has the honor of being the first synthetic ever, in 1913, to receive a patent. It typically is modified prior to processing, using

Figure 4.6 CD cases made from polystyrene. designelements/Shutterstock.com.

Figure 4.7 Audiophiles all over the world still enjoy listening to music on vinyl LP records. RG-vc/Shutterstock.com.

a wide variety of additives, such as stabilizers, plasticizers, and processing aids. The chemical and environmental resistance of a specific grade is dependent on the additives used. Its continued use is being questioned because many of these additives can leech out, causing environmental contamination or harm to humans.

Styrene acrylonitrile (SAN)

A copolymer, normally consisting of between 70–80% styrene and 20–30% acrylonitrile (another vinyl monomer), SAN is known for its barrier properties, weatherability, stress-crack resistance, toughness, and dimensional stability. Its make-up leads to higher strength, rigidity, and chemical resistance than PS, although it tends to yellow more quickly. A larger percentage of acrylonitrile adds a yellow tint, but also improves chemical resistance and mechanical properties. SAN has high smoke generation when burned, as well as a low heat-deflection temperature. It is less expensive than acrylic.

SAN is widely used. Industrial uses include dials, switches, lenses, knobs, and battery cases. Medical applications include dental and medical light diffusers. Within the electrical and electronics area, it is used for CD players, aid conditioner impellers, meter covers, telephone components, and radios. Among consumer products are toothbrush handles, cassette cases, disposable lighters, cosmetic containers, mixers, juicers, and clothes hangers.

Thermoplastic polyurethane **(TPU)**

A term used to describe a class of synthetic materials with very unique properties. Most TPU materials are *block copolymers*, meaning they consist of two or more homopolymers bonded together in a linear pattern (the term blocks refers to the distinct units that are bonded together). The properties depend on the pattern and ratio of the blocks. There are four primary kinds of TPU, involving two different chemical families (polyether and polyester) and two different chemistries (aliphatic and aromatic). Some versions can be processed as thermosets, others as thermoplastics. There are also a number of blends and alloys available.

There is wide variety in the properties of TPU, which are similar to those of thermoset polyurethanes, but they do not have the same level of heat resistance. They have high abrasion resistance, good property retention at low temperatures, good resistance to solvents, and good weatherability. In the rigid versions, TPUs are used in power tools, automotive instrument panels, medical devices, footwear, caster wheels, drive belts, and sporting goods, as well as in electronic equipment such as mobile phones (Figure 4.8).

An early use of TPU was as a replacement for PVC, with the goal of having the same look-and-feel, but with a much better performance in such areas as abrasion resistance, plasticizer migration, flex properties, and low-temperature toughness (Table 4.1, Figure 4.9).

Figure 4.8 Automotive dashboards are often made from TPU. Christian-Delbert/Shutterstock.com.

Table 4.1 A List of Commodity Plastics

Common Name	Chemical Name	Acronym	Trade Names
Acrylic	Polymethyl methacrylate	PMMA	Plexiglas
Polyethylene	Polyethylene	PE	Ferrene, Petrothene, Sanalite, Versadur, Unival
Low-density polyethylene		LDPE	Dowlex, Novapol
Linear low-density polyethylene		LLDPE	Dowlex, Novapol
Very low-density polyethylene		VLDPE	
Medium-density polyethylene		MDPE	Sclair
Cross-linked polyethylene		XLPE	Schulink

4: An Overview of Thermoplastic Materials

Table 4.1 A List of Commodity Plastics—cont'd

Common Name	Chemical Name	Acronym	Trade Names
High-density polyethylene		HDPE	Alathon, Eltex. Forar, Fortiflex, Hostalen, Marlex
High-density cross-linked polyethylene		HDXLPE	
High molecular weight polyethylene		HMWPE	
Ultrahigh molecular weight polyethylene		UHMWPE	Spectra, Dyneema, Lennite, Tivar, Duravar
Ultralow molecular weight polyethylene		ULMWPE	
Polypropylene	Polypropylene	PP	Ferrex, Versadur
Polystyrene (styrene)	Polystyrene	PS	Dylene, Dylite, Ladene, Novacor, Replay, tyrofoam, Valtra
SAN	Styrene acrylonitrile	SAN	Lustran, Tyril, Luran, Rovel
PVC	Polyvinyl chloride	PVC	Anwidur, Apex, Astavin, Geon, Palvinyl, Tecavinyl, Trovidur, Versadur
Thermoplastic polyurethane	Polyurethane	TPU	Desmopan, Irogran, Avalon, Krystalgran, Irostic

Figure 4.9 Some examples of products made from engineering plastics. (a) Acetal buckle (viphotos/Shutterstock.com), (b) nylon rope (design56/Shutterstock.com), (c) rotary dial telephone (MilosLuzanin/Shutterstock.com), (d) Lego blocks (ESOlex/Shutterstock.com), (e) nylon fishing line (AdOculos/Shutterstock.com), and (f) BMX bike wheel.

4.1.2 Engineering Plastics

The term engineering plastics is used to describe a category of thermoplastic materials. These materials typically have better mechanical properties than commodity plastics, and as such are often used for structural purposes. The term engineering is derived from the fact that their performance can be predicted using traditional engineering calculations. In addition, there is substantial property data available for these materials, accounting for both short- and long-term use under a variety of end-use conditions. Key engineering plastics include the following:

Acrylonitrile butadiene styrene (**ABS**)

A key engineering material, suitable for a wide variety of structural applications. An alloy made from three polymers, its name is derived from them: A: acrylonitrile (a vinyl monomer); B: butadiene (rubber); and S: styrene.

As with most alloys, ABS offers a balance of properties. It has good strength, good stiffness, and good toughness. Numerous grades are available, including those especially formulated for super high impact, with excellent low-temperature toughness. The material properties and the cost of raw materials vary depending on the grade, which usually is determined by the ratios of the three polymers. By default, ABS is a very high-gloss material because of the styrene component. Thus, it is an excellent choice for consumer applications and it frequently is used in housings for consumer electronics, consumer appliances, and general purpose power tools.

ABS is an amorphous thermoplastic, which means it has relatively low mold shrinkage; but, it also means that it is less resistant to chemical attack than a semicrystalline thermoplastic. The chemistry of ABS is such that it can easily be painted or even chrome plated.

ABS is one of the older engineering plastics, first commercialized by Borg Warner in the 1950s under the trade name Cycolac. One of the first applications was for the housings of household telephones made by Bell telephone. Remember those big, old, black housings? The specific grade used in those phones was given the name Cycolac T (with "T" for telephone). Borg Warner manufactured Cycolac until GE Plastics bought its ABS business in the late 1980s. SABIC Innovative Plastics then bought the GE Plastics business in 2007, and renamed it. So all the former GE materials (Xenoy, Lexan, Noryl, Cycolac, etc.), are now made by SABIC. SABIC is an international conglomerate and is now one of the world's largest suppliers of thermoplastic materials (Figure 4.10).

In summary, ABS is a workhorse material and should be considered a staple in the portfolio of any design engineer.

Acetal

Acetal is one of my favorite engineering materials. Also known as polyoxymethylene (POM), polyacetal, and polyformaldehyde, acetal is a fascinating material often overlooked by most design engineers.

Acetal is a semicrystalline thermoplastic with an interesting mix of properties. First, it has excellent mechanical properties, including high strength, high stiffness, and high toughness. Second, it is resistant to a wide variety of chemicals—a result of its crystalline structure. Third, acetal has unique characteristics in the area of tribology—the science of friction, lubricity, and wear. It has a very low coefficient of friction, and outstanding wear characteristics—and all of these characteristics

Figure 4.10 Classic rotary dial telephone. ARENACreative/Shutterstock.com.

can be enhanced even further with additives such as chemical lubricants and polytetrafluoroethylene (PTFE). Finally, acetal has some unique molding properties. Because it is a crystalline thermoplastic, it has a distinct melting point. This means that flow lengths in the mold are related to temperature and NOT to pressure. Acetal also has exceptional knit-line strength—in some instances, the knit-line strength can approach 100% of the base material. This means that you can design—and then mold—extremely complex parts with high structural integrity (just try doing that in ABS).

There are two main types of acetal: homopolymer and copolymer. The homopolymer version was invented by DuPont in the 1950s, with the trademark name of Delrin®. The copolymer version came later, under the brand names Celcon®, Hostaform®, Ultraform®, and Duracon®. The distinctions between these two versions are subtle:

- *Homopolymer* has better mechanical properties, including higher mechanical strength, higher hardness, greater stiffness, and better creep resistance. It also has a slightly higher melting point than copolymer.
- *Copolymer* has lower crystallinity (less shrinkage, better dimensional stability, but also different chemical resistance).

Other differences are as follows. Homopolymer is produced through the polymerization of formaldehyde. Upon decomposition, homopolymer will produce formaldehyde. While stabilizers are added to homopolymer resins to prevent decomposition during the normal molding cycle, there are sometimes "issues" that occur where molten resin can accumulate and decompose (hold-up spots in the barrel, mismatch between the nozzle and the sprue, etc.). In my days at DuPont, we used the phrase "If it smells, it tells." However, most molders believe their molding facility is perfect, and will never admit they have an "issue." As a result, many molders prefer to use copolymer.

In the real world, homopolymer and copolymer often are interchangeable. However, there are those who claim (and I am one of them), that you can ALWAYS injection mold a part faster with homopolymer than you can with a copolymer because the higher melting temperature means it will solidify faster than a copolymer during the injection molding process.

Acetal also is a great material for machining. As mentioned earlier, it is stiff, strong, and hard, and offers exceptional frictional properties. All of these properties combine to make it easy to machine. It therefore is great for making prototypes. It also has a very pleasing surface appearance, regardless of whether it is molded, extruded, or machined.

Typical applications for acetal include molded gears, mechanisms, disposable lighters, and snap-fit connectors. No doubt about it, acetal is a critical component of the material palette for any serious design engineer (Figure 4.11).

Note: Acetal, compared to other thermoplastic materials, does have a very high level of crystallinity (75–85%). This is what gives the material its strength and stiffness, and its resistance to chemical attack. It also means that molded acetal has a high amount of mold shrinkage. So, one needs to plan for this in making design and tooling decisions.

Polyamide (PA)

Invented in 1935 by DuPont chemist Wallace Carothers, PA are a family of polymers. The generic name for the family is "nylon." These polymers contain chains of monomers that are bonded together via amide bonds, hence the term "polyamide." Nylon can be fabricated as a copolymer, with two different types of monomer, or as a homopolymer using a single monomer. The type of nylon is designated by the number of carbon atoms in the base monomers. The most common types are nylon 6 and nylon 6/6, but there are also more specialized types, including nylon 11, nylon 12, nylon 5/10, and nylon 6/12.

Figure 4.11 Disposable lighters molded from acetal. StudioKIWI/Shutterstock.com.

Nylon has some unique characteristics. First, it is a semicrystalline material, and like most semicrystalline materials has excellent chemical resistance. Second, it has a relatively high melting point and thus is able to withstand high end-use temperatures. Third, it inherently has a great deal of lubricity, and excellent resistance to abrasion and wear. Fourth, nylon absorbs moisture, either from the atmosphere, or directly from contact with water, which affect the physical properties (making it tougher and more flexible), and also the overall dimensions. The amount of moisture in the material is dependent on the relative humidity.

Nylon also has good mechanical properties. It has good strength, and reasonable stiffness. It has good elongation, and a unique yield behavior under tension. Instead of undergoing brittle fracture, the material will continue to elongate and "neck down," and the polymer chains will align in the direction of elongation, increasing its tensile strength. This phenomenon allows nylon to be drawn into fibers with very high strength. These fibers are used for fabric, carpet fibers, rope, and fishing lines.

Below are some common nylon types.

- *Polyamide 6/6* (Nylon 6/6, PA 6/6) The semicrystalline PA is one of the most popular engineering thermoplastics; it is often used when high mechanical strength, great rigidity, and good stability under heat are required. It is widely used in the automotive industry for under-the-hood applications, such as fans, washers, brackets, and housings. With glass reinforcement, it is also used for radiator end caps and intake manifolds. Other popular applications include zip ties, carpet fibers, and sporting goods equipment.
- *Polyamide 6* (Nylon 6, PA 6) A semicrystalline PA with a slightly different chemistry than PA 6/6. It is unlike most nylons because it is formed by ring-opening polymerization rather than being a condensation polymer. PA 6 has a melting temperature that is slightly lower than that of PA 6/6, so it has slightly lower heat resistance. In injection molding, parts made from glass-reinforced PA 6 have a higher surface gloss than parts made from glass-reinforced PA 6/6. Because of this, many power tools and office furniture parts are made from glass-reinforced PA 6 (or from a glass-reinforced PA 6 / PA 6/6 alloy).

  The use of PA 6 range from toothbrush bristles to gears, fittings, and bearings. It is also used as surgical sutures, and strings for both classical and acoustic musical instruments ... and tire cords, which could be why you hear your new tires singing down the road. PA 6 fibers have luster and elasticity, as well as high tensile strength.
- *Polyamide 6/10* (Nylon 6/10, PA 6/10) A semicrystalline PA with low moisture absorption compared to other nylons. Thus, it retains its properties better when wet. Similar to most forms of nylon, PA 6/10 is smooth to the touch and can be manufactured to range in luster from dull to shiny. Nylon 6/10 can be extruded into thin filament fibers to make fabric fibers and brush bristles, or casted into a predetermined shape, similar to other hardened plastic parts. The filament can be dyed by infusing the plastic polymer with a coloring agent and the substance does not lose its color easily.
- *Polyamide 6/12* (Nylon 6/12, PA 6/12) Another low-moisture-absorbing nylon. Due to its chemistry, it also undergoes less dimensional change when moisture is absorbed. It has lower mold shrinkage than PA 6, PA 6/6, or PA 6/10, and better heat resistance than PA 6/10. It has a higher cost relative to other nylons. It is often used as a coating material.

- *Polyamide 11* (Nylon 11, PA 11) Another low moisture nylon, PA 11 is unique in that it is derived from an organic material—castor beans. It has a great balance of properties, and is also quite flexible. It is often used in hydraulic hoses, and in tubing for vacuum, fuel, vapor, air, and brake lines.
- *Polyamide 12* (Nylon 12, PA 12) Of all nylons, this has the lowest moisture absorption and is tougher than PA 11. PA 12 is the toughest and most flexible, in dry conditions, of all nylon. When saturated in water, it is stiffer and more rigid than other nylons. Whether wet or dry, it has the greatest resistance to abrasion. The crystallite melting point is lower than that of other nylons, giving nylon 12 the lowest heat distortion temperature and softening point. It resists high impact at low temperatures when dry, but nylon 6 and 6,6 have better impact resistance when wet.
- *Amorphous nylon* Copolymer PA that have no crystallinity in their polymer structure. These materials offer strength and performance similar to other nylon materials, while also providing high transparency (i.e., they are clear). Compared to other clear (amorphous) resins, they offer advantages such as excellent dimensional stability and good chemical resistance. Applications include displays, safety equipment, and medical devices.

Added together, PA offer an unparalleled combination of strength, stiffness, and elongation. One application that demonstrates this succinctly is a standard cable tie. The tail end of the tie provides the structure, and also has an angled gear rack molded into it. The head of the tie contains a flexible ratchet. This ratchet flexes to allow the tail to be inserted, and then flexes click-click-click over the gear rack as the tie is tightened. But, if the tail is pulled backward, the ratchet locks in place, and the tail cannot be removed (Figure 4.12).

Nylon chemistry also facilitates the use of specialty additives and modifiers. Grades of nylon are available with lubricants, impact modifiers, reinforcing agents, and processing agents. Glass fibers are commonly used as a reinforcing agent. Glass-reinforced nylon has become the material of choice for housings of high-performance power tools due to its combination of strength, stiffness, durability, and chemical resistance.

PA also have a unique chemical structure that accommodates the use of dyes and colorants. Custom colors easily are achieved in the injection molding process through master-batch concentrates, cube blending, or custom compounding.

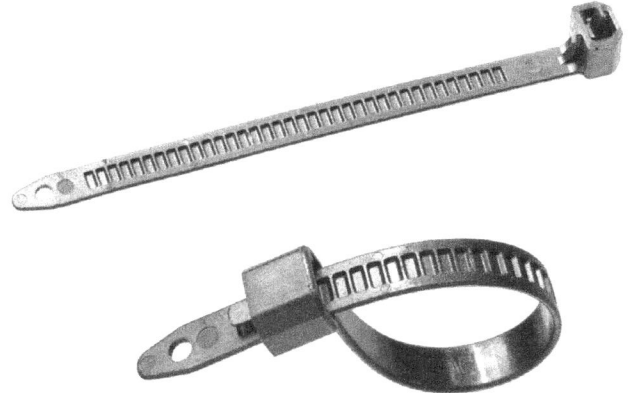

Figure 4.12 Cable tie. WinaiTepsuttinun/Shutterstock.com.

Nylon is used in high-performance applications in virtually every industry, including automotive, power tools, and office furniture (chair bases and chair shells). The grandfather of all plastics, nylon remains one of the world's most widely used engineering materials.

Polycarbonate

Polycarbonate is an amorphous thermoplastic with a fantastic balance of properties. It is stiff, strong, and has good ductility. But, in my mind, what makes polycarbonate unique is its exceptional impact resistance. While it does have some notch sensitivity (as do most thermoplastic materials), in solid or sheet form the impact resistance of polycarbonate is unmatched. As an example, thermoformed polycarbonate is used for the cockpit canopy of several fighter jets, including the F-22 Raptor. If your application involves any kind of impact, polycarbonate should be on your short list of materials to consider.

Polycarbonate was commercialized in the 1950s, "invented" at almost the same time by GE Plastics in the USA and by Bayer in Germany. GE marketed the material under the trademark Lexan®, while Bayer marketed the material under the trade name Makrolon®. Polycarbonate is used in a wide variety of applications, including appliance housings, football helmets, automotive headlights, and CD and DVD discs.

In addition to its physical properties, polycarbonate is also easy to process, and can be manufactured in sheets. In this form, it is widely used in the construction industry, especially for large windows and architectural lighting applications.

Its optical properties are similar to PMMA, but with slightly lower optical clarity. However, it does have a higher refractive index, so lenses in polycarbonate often can be made thinner than in other materials (including PMMA). In eyeglasses, this allows for thin, lightweight lenses. In addition, the exceptional impact resistance of polycarbonate is an added safety feature.

I often think of polycarbonate by comparing it to other materials.

"It's kind of like acrylic, but with better physical properties." (and it's more expensive)

"It has a balance of properties like ABS, but with better toughness." (and it can be made clear)

It has properties similar to nylon, but it is amorphous.

While none of these descriptions truly do it justice, they may help you better understand the unique characteristics of this material (Figure 4.13).

Figure 4.13 Most modern fighter jets have a canopy made from polycarbonate. VanderWolfImages/Shutterstock.com.

Polyester

Describes a number of polymers that contain an ester group in the polymer chain. This includes a wide range of materials, including organic and synthetic materials, and thermoset and thermoplastic resins. For design engineers, there are two important thermoplastic polyester resins: polyethylene terephthalate (PET) and polybutylene terephthalate (PBT).

- *Polyethylene terephthalate* Has a high melting temperature and exhibits excellent chemical resistance. It can exist in an amorphous or semicrystalline state, depending on how it is processed. The primary use of PET is in synthetic fibers for the textile industry (where it usually is described simply as "polyester").

The next largest use is in packaging applications, where it is used as a film or in rigid form for food containers, plastic jars and bottles, and beverage containers. PET packaging is easily recycled and is frequently reprocessed into carpet fibers, fleece jackets, and insulating fill. PET is the world's third most produced polymer (after polyethylene and polypropylene).

PET also is formulated as an engineering material for injection molding, usually with glass fiber reinforcement. First marketed by DuPont under the trade name Rynite®, glass reinforced PET offers exceptionally high strength and stiffness, outstanding chemical resistance, and high temperature performance.

Note: PET is NOT related to PE, and is more accurately described as *poly(ethylene terephthalate)*.

- *Polybutylene terephthalate* A semicrystalline engineering-grade thermoplastic. Its advantages are low water absorption, strength, and moldability. In comparison with PET, PBT has somewhat less stiffness and strength, but has higher impact strength and similar chemical resistance. It is affected by boiling water, and can be used only up to 300 °F. Because it crystallizes more rapidly than PET, it often is preferred for industrial scale molding.

  PBT is used for housings in electrical applications, in automotive construction as plug connectors, and in households in showerheads or irons. It is also found processed into fibers in toothbrushes and is used in the keycaps of some mechanical keyboards because of its resistance to wear.

  PBT can also be made into yarn. This has a natural stretch similar to spandex and can be incorporated into sportswear. Because of its chlorine resistance, it is found especially in swimwear.

- *Polycyclohexylenedimethylene terephthalate* A semicrystalline polymer that is similar to PET. It has many of the same characteristics, but has much higher heat resistance. It is often described as a high-heat polyester.

- *Polytrimethylene terephthalate* A semicrystalline polymer that is similar to PBT and has many of the same characteristics. First synthesized in 1941, it has seen little commercial application because it was more expensive to produce than either PBT or PET. However, recent developments—including the ability to generate the intermediate propanediol from corn sugar—have helped make it more cost effective. This is also a significant advance in the use of renewable materials.

Thermoplastic polyesters are demanding materials to mold. They require special attention to keep the material dry (both before and during the molding process), as well as special attention to the temperature profile of the melt from the barrel to the mold. The design of the injection mold itself also is important because heat transfer in the mold is critical. The properties of the molded part are affected by the crystallization of the material, and this crystallization is dependent on the heat transfer dynamics of the molding process.

In a final molded part, PET and PBT have exceptionally low moisture absorption and a low coefficient of thermal expansion. This gives them excellent dimensional stability over a broad temperature range. When proper molding controls are implemented, PET and PBT parts can be molded with exceptional precision. This makes them excellent candidates for miniature electronic components.

PET and PBT frequently are used as a replacement for die-cast metal parts. They are also used extensively in electronics applications, both as a replacement for epoxies and other thermoset materials, as well as for housings and plugs and receptacles. They are also used as housings in high-temperature applications: clothing irons, hot-melt glue guns, etc. (Figure 4.14).

Polyphenylene oxide (**PPO**)

An amorphous thermoplastic consisting of phenylene monomers linked together via ether bonds. There are actually two classes of

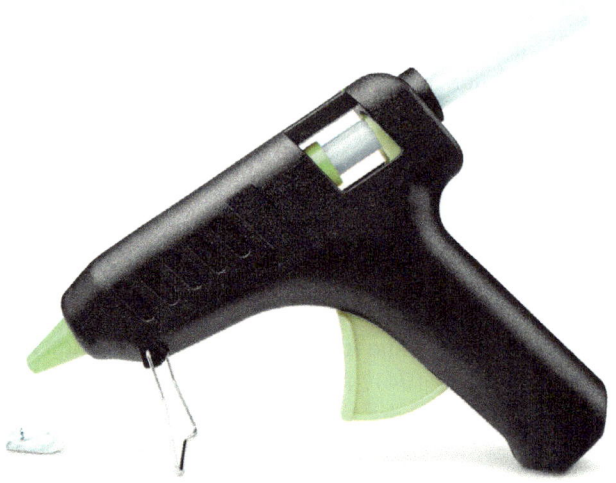

Figure 4.14 Hot melt glue gun. NilsZ/Shutterstock.com.

polyphenyl materials, depending on the chemistry of the ether bond: one is PPO, the other is polyphenylene ether. While PPO by itself is unsuitable for injection molding, it is easily blended with other materials (including PS, ABS, nylon, or thermoplastic polyester) to improve processing, reduce cost, and target-specific performance requirements.

GE Plastics first introduced a modified PPO in the early 1960s under the trade name Noryl®. Noryl was a blend of PPO and PS, and the properties and characteristics of each grade were dependent on the ratio of the blend. Unlike a number of other commercial polymers—which were alloys of different polymers or copolymers—Noryl® was a true blend, where the materials were simply mixed together with no chemical bonding between the base polymers. The blending capability of PPO is an important characteristic to remember.

PPO resins have an excellent blend of physical properties, along with excellent electrical properties. PPO also has an exceptionally high glass transition temperature. As a result, parts made of PPO have a very high useful temperature range. Some grades often are used in medical applications for parts and components that must undergo sterilization (via steam, gamma radiation, or ethylene oxide). PPO blends are also used for structural parts, electronics, household appliances, and automotive components.

Taken as a class, modified PPO is one of the big five engineering plastics (along with nylon, polycarbonate, acetal, and thermoplastic polyester). Modified PPO is an important engineering material and should be on every design engineer's palette of materials (Table 4.2, Figure 4.15).

Table 4.2 A List of Engineering Plastics

Common Name	Chemical Name	Acronym	Trade Names
ABS	Acrylonitrile butadiene styrene	ABS	Cycolac, Absylux, Lustran, Magnum, Novodur, Royalite, Teluran, Toyolac
Acetal	Polyacetal, polyoxymethylene	POM	Delrin, Celcon, Duracon, Hostaform

Continued

Table 4.2 A List of Engineering Plastics—cont'd

Common Name	Chemical Name	Acronym	Trade Names
Nylon	Polyamide	PA	Zytel, Durethan, Ultramid
Amorphous nylon		PA	
Nylon 11		PA 11	
Nylon 12		PA 12	
Nylon 6		PA 6	
Nylon 6/10		PA 6/10	
Nylon 6/12		PA 6/12	
Nylon 6/6		PA 6/6	
High-temperature nylon		HTN	
Polycarbonate	Polycarbonate	PC	Lexan, Makrolon, Calibre
Polyester			
Polybutylene terephthalate	Polybutylene terephthalate	PBT	Valox, Crastin, Duranex, Ultradur
Polyethylene terephthalate	Polyethlyene terepthalate	PET	Rynite, Mylar, Unitep, Kodapak, Selar
PCT	Polycyclohexylenedimethylene terephthalate	PCT	Ektar
Polytrimethylene terephthalate	Polytrimethylene terephthalate	PTT	Sorona, Tritan, Corterra
Modified polyphenylene oxide	Polyphenylene oxide	PPO	Noryl, Norylux, Uninor, Iupiace

4: An Overview of Thermoplastic Materials

Figure 4.15 Some examples of products made from specialty plastics. (a) Kevlar vest (USAartstudio/Shutterstock.com), (b) Nomex firesuit (NikitinVictor/Shutterstock.com), (c) teflon pans (t81/Shutterstock.com), (d) LCP electronic connector (30-pin) (DenisDryashkin/Shutterstock.com), (e) flex circuit cable (ZigaCetrtic/Shutterstock.com), and (f) Kevlar speaker (AleksandarPulios/Shutterstock.com).

4.1.3 Specialty Plastics

The term specialty plastics is used to describe a category of thermoplastic materials with significantly higher performance characteristics than traditional engineering materials. They are usually produced in much lower quantities than commodity or engineering plastics, at a higher cost. Most materials in this category have unique properties in one or more areas: they may be impervious to chemicals or capable of withstanding extreme environmental conditions. Key specialty plastics are discussed.

Aramid

The term aramid is a portmanteau of the words *aromatic polyamide*. PA, as you may recall, is the chemical name for the family of materials known as nylon. In organic chemistry, the term aromatic refers not to the way something smells, but to the way in which molecules bond in a ring form, creating strong and stable molecular bonds that are often superior to molecular bonds made in a linear fashion. In polymer chemistry, aromatic polymers typically have performance advantages over linear or cross-linked polymers. This is especially true for aramids (Figures 4.16 and 4.17).

Aramid molecules are particularly well suited to be made into synthetic fiber. Some well-known aramid fibers include Nomex® and Kevlar®, both developed by DuPont. Nomex® has exceptionally high heat resistance and is often used to make flame-proof clothing. Kevlar® has exceptionally high strength, with a tensile strength-to-weight ratio five times greater than steel.

Nomex® can be made in both a fiber and paper form. In paper form it is often used for electrical insulation, or to make laminated fire-proof structures. In a fiber form, it is often woven into fire-proof fabric, which is then used to make clothing worn by firefighters, pilots, and automobile racing drivers. Military tank drivers often wear Nomex® hoods to protect against both fire and extreme cold. Military pilots and aircrew often wear flight suits made from Nomex®, along with other fire-retardant materials. Racing drivers and crews also wear clothing made from Nomex® for similar protection: gloves, long underwear, balaclavas, socks, helmet lining, and shoes.

Kevlar® fiber can also be used to make fabric or woven into rope or twine. In fabric form the high tensile strength of the individual fibers results in a fabric that is exceptionally resistant to tearing or puncture. Fabric made from Kevlar® fiber is often used to manufacture gloves, sleeves, jackets, chaps, and other clothing designed to protect

4: An Overview of Thermoplastic Materials

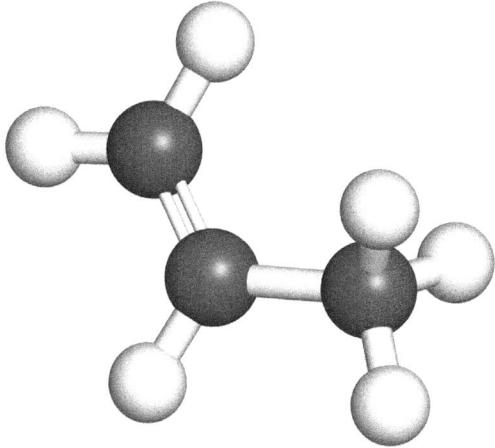

Figure 4.16 A monomer of polypropylene, showing aliphatic bonds. molekuul.be/Shutterstock.com.

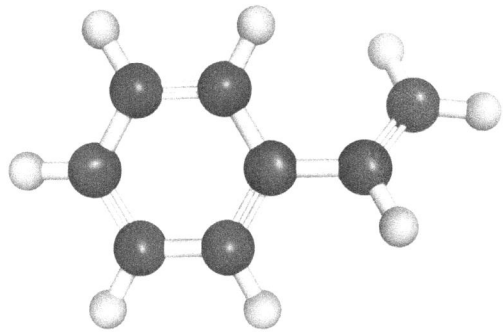

Figure 4.17 A monomer of styrene, showing an aromatic bond. Iculig/Shutterstock.com.

users from cuts, abrasions, and heat. It is perhaps best known for its use as a "bullet-proof" fabric, where layers of woven fabric are used to create body armor such as helmets, face masks, and vests.

Kevlar® fiber is also used as a reinforcing agent in thermoset composite structures and in thermoplastic resins. Its high strength-to-weight ratio allows manufacturers to build lightweight products with exceptional strength. As such, it is popular in the sporting goods industry. Tennis racquets are often strung with Kevlar® fiber, and it is used to make light and durable sails for high-performance racing boats (Figure 4.18).

Figure 4.18 Speaker cone fabricated from woven aramid fibers. sofist/ Shutterstock.com.

Fluoropolymers

Another unique family of specialty materials. They are the result of having one or more hydrogen atoms in the polymer chain replaced by fluorine atoms. The resulting polymer can be described as being partially fluorinated or fully fluorinated.

Fluoropolymers typically have good mechanical properties, with the added advantage of high temperature performance coupled with outstanding chemical resistance. Most of them also have a very low coefficient of friction, sometimes so low that it cannot be believed. There are numerous varieties of fluoropolymers. They can be used as fibers, as films, as powders, or in a solid form. Fluoropolymers play a crucial role in many products and services on which we depend on in our daily lives—yet most of us do not realize it. Some of the more commonly used fluoropolymers are described below.

Polytetrafluoroethylene **(PTFE)**

PTFE was an accidental invention made in 1938 by DuPont chemist Roy J. Plunkett. Tests revealed that this new material was not only

Figure 4.19 Thread sealing tape made from PTFE film. DmitryKramar/Shutterstock.com.

slippery to the touch, but it had exceptional high-temperature stability and was also resistant to corrosion from most substances. It was later given the trademark Teflon®. An early use of PTFE was in the Manhattan Project as a coating material to protect the piping used in uranium enrichment. It soon found its way into the home, where it was used as nonstick coating on cooking pans.

Today, PTFE is widely used to coat wiring in aerospace and computer applications (e.g., hookup wire, coaxial cables), primarily because of its excellent dielectric properties. Combined with its high melting temperature, it is often the material of choice as a high-performance substitute for the weaker and lower-melting point PE and PVC coatings (Figure 4.19).

Ethylene tetrafluoroethylene (**ETFE**)

Similar to PTFE, ETFE has a higher tensile strength, but less toughness and an even lower coefficient of friction. It also has different chemical and oxidative properties, and is lower in cost.

ETFE is commonly used in the nuclear industry for tie or cable wraps, and in the aviation and aerospace industries for wire coatings. It can be made into sheet form that is strong and elastic, and frequently used in architecture. Thin ETFE sheets were used to cover the Beijing National Aquatics Centre that hosted the 2008 Olympic swimming events.

Perfluoroalkoxy alkane (PFA)

A fully fluorinated fluoropolymer, PFA has a slightly different chemistry than PTFE, where alkoxy groups (carbon hydrogen chains) replace some of the fluorine atoms. While the properties are similar to PTFE, the biggest difference is that PFA can be melt-processed (molded, extruded, etc).

Fluorinated ethylene propylene (FEP)

Another fully fluorinated fluoropolymer, FEP is a copolymer of hexafluoropropylene and tetrafluoroethylene (the base material of PTFE). It was the first commercially produced material that offered the unique advantages of fluoropolymers with the melt-processing capabilities of more conventional polymers. Properties are very similar to PTFE, with a lower service temperature.

Polyvinylidene fluoride (PVDF)

Another melt-processible fluoropolymer, PVDF is known for its abrasion resistance and rigidity, and lower cost (when compared to other fluoropolymers). It is virtually unaffected by long-term exposure to UV radiation. Inherently flame retardant, it is classified as "Self Extinguishing, Group 1" by Underwriters Laboratories, Inc.

Liquid crystal polymer (LCP)

Liquid-crystal polymers are a family of materials based on aromatic polyester chemistry. Like other aromatic polymers (such as aramids), the aromatic bonds provide them with enhanced performance over traditional polyester materials. The term *liquid crystal* implies that they have characteristics of both liquid and solid (e.g., crystalline) materials. (And you have to admit the phrase *Liquid Crystal Polymer* sounds much better than the term *Aromester*.)

In a fiber form, LCPs are used in a similar manner as aramid fibers. In solid form, as is used in molding and extrusion, or even with thermoforming, LCPs are quite unique. In the molten phase, the molecular chains are oriented and aligned, without entanglements. This means the molten material can flow easily (like a liquid), allowing for fill of very thin sections at very low pressure. They then solidify very quickly allowing for very fast molding cycles on parts with incredible detail and precision. Unlike many other specialty plastics, they can often be processed on conventional equipment. This ease of forming can be an important competitive advantage against other plastics, as it offsets the high raw material cost.

As a family, LCPs have high strength, high stiffness, excellent chemical resistance, and they are extremely stable at high temperatures. They also have inherent flame retardancy and good UV resistance (which usually results in good weatherability). LCPs are frequently used in electrical applications (such as bobbins, sockets, and connectors) and as a replacement for stainless steel in medical applications (Figure 4.20).

Polyarylate **(PAR)**

Another family of aromatic polyesters. However, unlike LCP (the other aromatic polyester), they are only partly aromatic and are based on a completely different chemistry. The chemistry is closer to that of polycarbonate, in that the PAR molecules are linked together by ester groups (polycarbonate utilizes an ester made from carbonic acid).

PAR have similar properties as polycarbonates including high strength, good toughness, and good clarity. They have a higher melting point, so offer better high-temperature performance. Other PAR characteristics include chemical resistance, good electrical properties, excellent flexural recovery, and exceptional UV stability. They are higher priced than polycarbonate.

PAR are employed in automotive parts, semiconductor components, ovenware, electronic devices, solar energy components, appliance parts, and snap lock connectors.

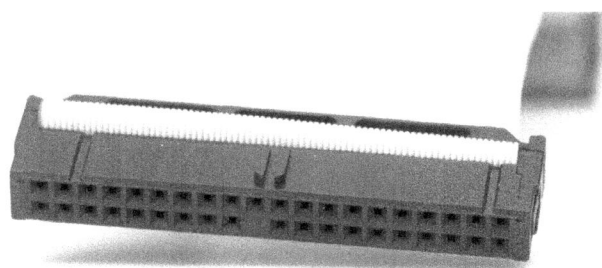

Figure 4.20 Multipin connectors are often molded from liquid crystal polymer. ravl/Shutterstock.com.

Polyimide (**PI**)

A family of materials where the polymer molecule consists of a chain of imide monomers. The links in these chains can be linear (also called aliphatic), in a ring form (aromatic), or a combination (semiaromatic). Aromatic polyimides (*arimides*?) are the most widely used.

PI can be either thermoset or thermoplastic. Early thermoplastic versions were not completely melt processible. Subsequent development led to improved versions that could be injection molded and/or extruded. These versions are often referred to as *thermoplastic polyimides* (TPI).

PI are extremely strong and heat- and chemical-resistant polymers and are also resistant to wear and radiation. In a thermoset form, PI are frequently used in a film shape, either from a roll or sheet. Due to their strength, they can be made extremely thin, and as a result are lightweight and flexible. They are often used for flexible cables, in laptops, cell phones, and other electronics applications (Figure 4.21).

As a thermoplastic, PI can be used in a film or solid form. Most TPI versions are semicrystalline, however, amorphous versions are also available. In either case, TPI is one of the most heat-resistant thermoplastics available. It has excellent physical properties, along with chemical and wear resistance. TPI is also an inherent flame retardant. TPI applications include thrust washers and seal rings for automotive and off-road vehicle transmissions, thermal insulators and stripper fingers for high-speed copiers, jet engine components, check valve

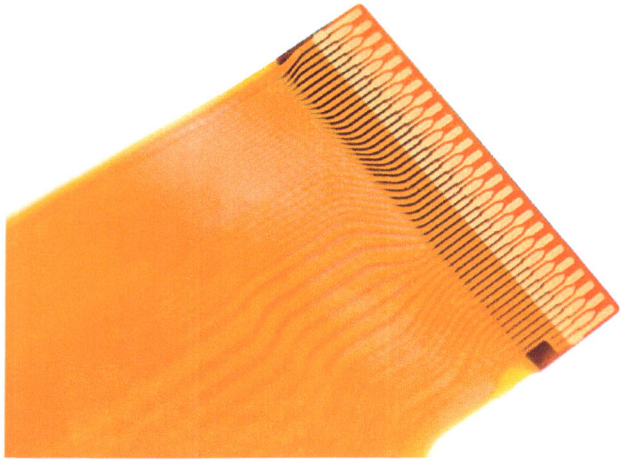

Figure 4.21 Flexible printed circuits are frequently made from polyimide film. ZigaCetrtic/Shutterstock.com.

balls, spline couplings, heat-resistant gears, vanes, wear strips, valve seats, journal bearings, and bearing retainers. Their strength and other characteristics make them ideal for use in automobiles, from chassis and struts to some parts found under the hood. Resistance to lubricants, fuels, and coolants are also major pluses.

There are a few additional families of materials with chemistry that is similar to PI, but with minor differences. These families include polyamide-imide (PAI) and polyetherimide (PEI).

Polyamide-imide

PAIs are unique materials that have elements of PA (aka nylon) chemistry, as well as aromatic polyimide chemistry. They have exceptional mechanical, chemical, and thermal properties and are considered by some to be at the top of the thermoplastic performance chart. They have high strength, exceptional high heat capability, and broad chemical resistance. Polyamide-imide polymers are melt processable and can be processed into a wide variety of forms—from injection- or compression-molded parts and ingots—to coatings, films, fibers, and adhesives. PAI is often lower in cost than TPI.

Parts molded from the PAI can maintain structural integrity in continuous use at temperatures up to 500 °F. Extremely resistant to flame, they have very low smoke generation. PAI is well suited for applications in extreme service conditions. PAI is often used in and around automotive engines, for gears, bushings, washers, and seals.

Polyetherimide

An amorphous thermoplastic with both ether links and imide groups in its polymer chain. First introduced by General Electric under the trade name Ultem®, they had discovered that the ether links allowed the material to be melt processable, while retaining many of the performance advantages of PIs. As a result PEI can be thermoformed, extrusion blow and injection molded, and/or extruded, although high processing temperatures are required.

PEI is known for long-term heat resistance, dimensional stability, and excellent stability of physical and mechanical properties at elevated temperatures. It has predictable stiffness and strength up to 200 °C. It also has inherent flame resistance and is difficult to ignite. PEI is used in medical and chemical instrumentation because of their heat resistance, solvent resistance, and flame resistance. It is also used in electrical/electronic applications, and automotive and aerospace applications.

Polyketone (PK)

A family of materials featuring ketone groups in the polymer chain. The links in these chains can be linear (aliphatic), in a ring form (aromatic), or a combination (semiaromatic). Aromatic polyketones are the most widely used. PK are another unique family with exceptional high-temperature properties, although the word itself sounds like a name of a 1950s doo-wop group, *"Ladies and gentlemen, let me present tonight's featured performers, The Ketones!"*

There are numerous materials in the PK family, including polyetherketone, polyetheretherketone, polyetherketoneketone, and polyetherketoneetherketoneketone.

While there are subtle differences within the family, all of these materials are semicrystalline engineering thermoplastics, with high strength and toughness, high chemical resistance, and high-temperature resistance.

Polymethylpentene (PMP)

A thermoplastic polymer of methylpentene monomer units. Although it is a semicrystalline material, it is transparent, with excellent transmission of visible light. It also has good transmission in the UV and radio wavelengths. It has an exceptionally low density (0.83 g/cc—the lowest of any thermoplastic), so parts molded from PMP will be lighter. Other features include chemical resistance, good thermal stability, toughness, medium heat resistance, and high rigidity. It has nearly no odor, low odor transfer, and it is acceptable for food contact.

Although high in cost, it is often used in areas where transparency is needed—in combination with thermal and chemical demands—such as autoclavable medical and laboratory equipment, microwave components, and cookware. Other applications include sonar covers, speaker cones, ultrasonic transducer heads, and lightweight structural parts. It is also FDA compliant for use in food processing machinery (Figure 4.22).

Polyphenylene sulfide (PPS)

A specialty engineering thermoplastic with a unique balance of properties. It provides high stiffness and strength at elevated temperatures, along with exceptional chemical resistance. It is a semicrystalline thermoplastic. Like all semicrystalline thermoplastics, it has both amorphous and crystalline components in its solid form.

PPS has outstanding chemical resistance, thermal stability, dimensional stability, and fire resistance. Its extreme inertness toward organic

Figure 4.22 Laboratory glassware.

solvents and inorganic salts and bases leads to outstanding performance as a corrosion-resistant coating suitable for contact with foods.

Phillips Petroleum introduced the first commercial PPS grades in 1968 under the trade name of Ryton. It had a very high viscosity and was initially a linear polymer that could be compression molded. Five years later, in Europe, the emphasis was shifted to coating processes and injection molding. It currently is used to make to make filter fabric for coal boilers, papermaking felts, electrical insulation, specialty membranes, gaskets, and various packaging.

Polyphthalamide (PPA)

A variation of the PA family (nylon), where a portion of the polymer chain is replaced with an aromatic component. Unlike aramids, which are fully aromatic polyamides, PPA materials are best described as being partially aromatic polyamides. The aromatic component extends the performance characteristics of the materials. They are often referred to as high-performance PA or high-temperature nylon.

There are numerous variations of PPA, just as there are with PA. In general, each variation of PPA extends the performance level of a PA with same carbon chain backbone resulting in better chemical resistance, higher heat resistance, and higher tensile strength.

Figure 4.23 Intake manifold.

PPA is extensively used in the automotive industry, especially for under-the-hood applications, including encapsulated solenoids, fuse holders, and brackets. They are also used for fuel and coolant lines, LED headlights, and various fittings and connectors.

Medical uses include tubing for devices such as catheters. Electronic applications are LEDs and cable/wire protection. PPA is used in gas pipes and supply lines in the oil industry, due to its ability to withstand high pressures.

PPA also offers direct bonding to many elastomers to provide plastic–rubber composites, either via overmolding or two-shot molding (Figure 4.23).

Polysulfone (PSU)

PSU are a family of materials with the sulfone monomer in the polymer chain. They are amorphous polymers, and are known for their toughness and stability at high temperatures. They have excellent strength and transparency, can withstand exposure to water—even at high temperature—and can be colored.

PSU has one of the highest service temperatures of all melt-processible thermoplastics. Its high hydrolysis stability allows its use in medical applications requiring autoclave and steam sterilization. However, it has low resistance to some solvents. PSU can also be reinforced with glass fibers, with a substantial increase in strength and stiffness. PSU are typically used in specialty applications or as a superior replacement for polycarbonates. One use is in structural applications where high heat and steam (autoclaving) is utilized to sterilize parts (such as with medical instruments).

4: An Overview of Thermoplastic Materials

There are a few additional families of materials with chemistry that is similar to PSU, but with minor differences. These families include polyethersulfone, and polyphenylsulfone.

Polyethersulfone

A variation in the chemistry with both sulfone and ether links in the polymer chain. The sulfone unit provides for thermal stability and strength, while the ether link provides improved processing (ease of molding).

Polyphenylsulfone

Combines high-performance properties such as excellent thermal stability, toughness, impact resistance, hydrolysis resistance, and stress-cracking resistance. It typically is used in the aerospace and automotive industries and in rapid prototyping and rapid manufacturing. It has outstanding color, transparency, and impact strength up to 200°C. PPSU has the best overall chemical resistance of the commercial PSUs.

Ultrahigh molecular weight polyethylene

UHMWPE is a term used to describe a version of PE with exceptionally long polymer chains. The longer chains provide stronger intermolecular bonds, which improves its chemical resistance and its mechanical properties, especially with regard to toughness. UHMWPE has the best impact performance of any thermoplastic. It also has a very low coefficient of friction, significantly lower than that of nylon and acetal, and comparable to that of PTFE. UHMWPE also has outstanding abrasion resistance that is better than PTFE, acetal, or nylon.

UHMWPE cannot be injection molded. It can be processed into solid blocks or fibers. In solid form (what is commonly known as the stock shapes industry), it can be available as blocks, sheets, rods, and tubes, which can then be machined into a final shape. It is used for bumpers, and rollers, and slides, and has even been used as synthetic ice in skating rinks.

In fiber form, the molecular chairs are aligned along the length of the fiber, providing exceptional tensile strength. UHMWPE fibers are also used in personal armor (and on occasion as vehicle armor), for cut-resistant gloves, climbing equipment, fishing line, and for high-performance sails and rigging in yachting. UHMWPE fibers excel as fishing line because they have less stretch, are more abrasion-resistant, and are thinner than traditional monofilament line.

Of all the various types of PE, my favorite is UHMWPE (or as I like to call it, oomm-whoopie). How can you not like a plastic called oomm-whoopie? (Table 4.3, Figure 4.24)

Table 4.3 A list of Specialty Plastics

Common Name	Chemical Name	Acronym	Trade Names
Aramid	Polyphenylene terephthalamides	PPTA	Kevlar, Technora, Twaron
	Polymeta-phenylene isophthalamide	MPIA	Nomex
Liquid crystal polymer	Aromatic polyester	LCP	Xydar, Zenite, Vectra
Polyarylate	Polyarylate	PAR	Ardel, Durel
Polymethylpentene	Polymethylpentene	PMP	TPX, Crystalor
Polyphenylene sulfide	Polyphenylene sulfide	PPS	Ryton, Supec, Fortron
High-performance polyamide	Polyphthalamide	HPPA/PPA/HTN	Amodel, Kalix
Fluoropolymers			
PTFE	Polytetrafluoro-ethylene	PTFE	Teflon, Neoflon, Hyflon
ETFE	Ethylene tetra-fluoroethylene	ETFE	Tefzel
Perfluoroalkoxy	Perfluoroalkoxy alkane	PFA	Teflon, Neoflon, Hyflon
Fluorinated ethylene propylene	Fluorinated ethylene propylene	FEP	Teflon, Neoflon, Hyflon
Polyvinylidene fluoride	Polyvinylidene fluoride	PVDF	Solef, Hylar, Kynar
Polyimides			
Thermoplastic polyimide	Thermoplastic polyimide	TPI	Kapton, Vespel
Polyamide-imide	Polyamide-imide	PAI	Torlon
Polyetherimide	Polyetherimide	PEI	Ultem

4: An Overview of Thermoplastic Materials

Table 4.3 A list of Specialty Plastics—cont'd

Common Name	Chemical Name	Acronym	Trade Names
Polyketones			
Polyetherketone	Polyetherketone	PEK	
Polyetheretherketone	Polyetheretherketone	PEEK	
Polyetherketoneketone	Polyetherketoneketone	PEKK	
Polyetherketoneetherketoneketone	Polyetherketoneetherketoneketone	PEKEKK	
Polysulfones			
Polysulfone	Polysulfone	PSU	Udel, Ultrason S
Polyethersulfone	Polyethersulfone	PES/PESU	Ultrason E
Polyphenylsulfone	Polyphenylsulfone	PPSF/PPSU	Radel

Figure 4.24 Collage of items made from TPEs.

4.2 Thermoplastic Elastomers

The term thermoplastic elastomer is used to describe another unique category of materials. Elastomers—from the words *elastic* and *polymer*—are materials that are inherently stretchy, or rubber-like. Rubber is a naturally occurring material (the sap from a gum rubber tree), and while there are synthetic rubbers, most of these are thermosets in nature. However,

TPE—just like other thermoplastic resins—can be repeatedly heated and melted and reprocessed. There are several different types of TPE, based on different chemistries. Below is a description of the different chemistries, arranged by their popularity.

Styrenic block copolymers (SBC)

Largest-volume category of TPE, with an annual consumption of about 1,200,000 metric tons. SBCs consist of at least three blocks: two hard PS end blocks and one soft, elastomeric midblock (either polybutadiene or polyisoprene). Polybutadiene and polyisoprene are synthetic rubbers and thermosetting in nature. However, the PS end blocks in the coploymer allow SBC materials to be processed as thermoplastics. It is essential that the hard and soft blocks are immiscible so that, on a microscopic scale, the PS blocks form separate domains in the rubber matrix thereby providing physical cross-links to the rubber.

Discovered and commercialized in the early 1960s, SBCs since have found numerous applications. Their typical balance between properties and processability leads to a focus on unique applications instead of a replacement for general purpose rubber. SBCs can be readily mixed with other polymers, oils, and fillers, which allows versatile tuning of product properties. They are employed in enhancing the performance of bitumen in road paving and roofing applications, particularly under extreme weather conditions. They are widely applied in adhesives, sealants, coatings, and in footwear. Also, SBCs are compounded to produce materials that enhance grip, feel, and appearance in applications such as toys, automotive, personal hygiene, and packaging. Kraton® is one widely recognized trade name.

Thermoplastic polyurethane

TPU elastomers are also block copolymers. There are four primary kinds of TPU, involving two different chemical families (polyether and polyester) and two different chemistries (aliphatic and aromatic). There are also a number of blends and alloys available. The properties depend on the pattern and ratio of the blocks.

TPU elastomers were the first major elastomers to be processed by thermoplastic methods. TPU elastomers are commonly used in inflatable rafts, swim fins and goggles, swimming pool cleaning devices, and drive belts.

TPUs can be processed on extrusion as well as injection, blow, and compression molding equipment. They can be solution-coated or vacuum-formed and are well suited for a wide variety of fabrication methods and configurations, including solids, films, membranes, and foams. TPUs can also be colored through a number of processes. But, more so than any other TPE, TPU can provide a number of physical property combinations making it a versatile material adaptable to dozens of uses:

- Abrasion resistance;
- Low-temperature performance;
- Mechanical properties, combined with a rubber-like elasticity;
- Shear strength;
- Elasticity;
- Transparency; and
- Oil and grease resistance (Figure 4.25).

Polyolefin blend elastomers (POE)

These materials consist of a polyolefin material (such as PE, PP, or polybutylene), or a copolymer based on a polyolefin, that is then blended with an elastomeric material (such as EPDM or vulcanized rubber) to create a flexible material that can be melt processed. They are sometimes also called thermoplastic vulcanizates (TPV).

If we go back to our analogy of baking a cake, thermoplastic vulcanizates are easy to understand. As you make the batter, you start with a mix of powdered flour. Think of this as the elastomeric material. You then add a liquid component—milk, and eggs, maybe some butter and

Figure 4.25 iDive housing.

vanilla flavoring. These represent the polyolefin in its molten state. You mix everything around, then pour the batter into a pan, and stick it in the oven. You let it bake (solidify), and then cool—and voila, you have a cake. The powdered flour is now suspended in a cooked medium, and the resulting cake is moist and flexible. If the cooked component of that cake batter could be turned back into liquid, you would have the culinary equivalent of a thermoplastic vulcanizate.

The POE/TPV family represents a relatively new class of polymers. They bridge the gap between impact-modified polyolefins (such as toughened PP) and other TPE. They are available in a broad range of hardnesses, with varying levels of stiffness and elongation. As polyolefins, they have excellent chemical resistance to acids, bases, and other aqueous media. Exposure to hydrocarbons can cause swelling due the EPDM component.

POE/TPV materials have excellent flex fatigue properties, good impact resistance, and they are low in cost and easy to process. They can be formulated to provide a soft, silky feel, which lends itself to portable electronic applications, such as smartphones, headphones, and portable computers. They are also used for both interior and exterior automotive parts, building materials, appliance parts, and sporting goods.

Limitations include moderate heat resistance, relatively poor wear resistance, and exposure to hydrocarbons causing swelling. Santoprene® is a widely recognized trade name.

Elastomeric Alloys

Elastomeric alloys are generated from a chemical combination of two or more polymers to give an alloy having better elastomeric properties than those of the corresponding blend. Unlike a blend—which is simply a mix of two or more items—the materials in an alloy will form chemical bonds between the different molecules, resulting in a new material with superior properties to any of the component materials. Elastomeric alloys are a new class of polymers that offer opportunities to explore entirely new areas of application.

Polyester block copolymers (PBC)

PBCs typically consist of two blocks—one being a hard polyester block, the other being a soft polyether block. They are also sometimes called thermoplastic polyester elastomers or thermoplastic polyether-ester elastomers.

PBC elastomers are high-performance engineering materials with exceptional properties. The polyester backbone provides for high

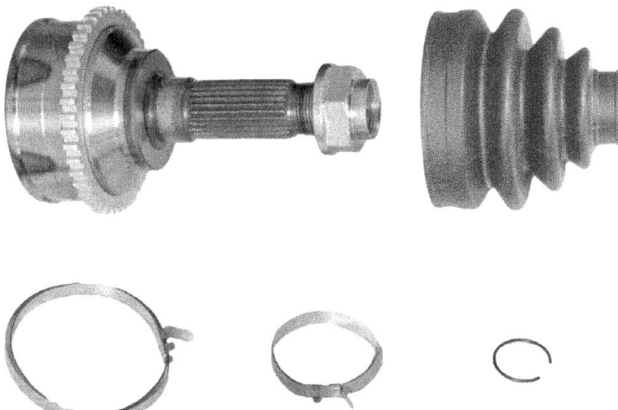

Figure 4.26 CV boot.

strength, good chemical resistance, impact resistance, high flex fatigue life (basically off the chart performance in bending), high tear strength, and flexibility at low temperatures. The measured value of any specific property (i.e., flexural modulus) depends on the composition and ratio of the polyester and polyether blocks.

PBC materials are melt processible and can be easily formed into sheets, films, or solid parts, using extrusion, injection and blow molding, melt casting, thermoforming, and more. Two widely known trade names are Hytrel® and Lomod®. In the automotive industry, the flexible boot used to protect the constant-velocity (CV) joint is often made from a PBC (Figure 4.26).

There are a few additional families of block copolymer elastomers with similar chemistry, but with minor differences. These families include polyamide elastomers (PAE) and polyether-block-amide elastomers (PEBA).

Polyamide elastomers

High-performance TPE based on a polyamide (aka nylon) hard block along with a soft segment. PAEs are also known as polyether block amides or as copolyester amides or thermoplastic elastomer-amides. They are used mainly in areas where other TPE cannot compete/perform, especially at lower temperatures.

PAE is used in such sports shoes as ski boots and soccer shoe soles, noiseless gears, and various types of hoses/tubings, such as paint spray hoses and vacuum brake booster lines.

Polyether-block-amide elastomer—Another high-performance TPE, featuring an amide (aka nylon) hard block instead of a polyester block, along with a polyether soft block.

PEBA is found in sports equipment, where it is used in damping systems and in the outsoles of high end shoes. It is used in winter sports gear because it enables the design of the lightweight alpine and Nordic ski boots while providing resistance to extreme environmental conditions.

In the medical field, PEBA is used as catheters because of its flexibility. It also is used in the manufacture of electric and electronic goods such as cables and wire coatings, electronic device casings, and components(Table 4.4).

Table 4.4 A list of Thermoplastic Elastomers

Common Name	Chemical Name	Acronym	Trade Names
Styrenic block copolymer elastomer	Styrenic block copolymer elastomer	SBC/ TPE-S	Kraton
Thermoplastic polyurethane elastomer	Thermoplastic polyurethane elastomer	TPE-U/ TPU	Estan, Elastollan, Pellethane
Thermoplastic polyolefin elastomer/ polyolefin blend elastomer	Thermoplastic vulcanizate elastomer	TPV	TPR, Santoprene, Geolast, Vyram, Nexprene, Alcryn
Polyester block copolymer elastomer	Thermoplastic polyether-ester elastomer	COPE/ TPE-E/ TPEE	Hytrel, Lomod, Arnitel, Riteflex
Polyamide elastomer	Polyamide elastomer	PAE	Vestamid, Grilamid
Polyether-block-amide elastomer	Polyether-block-amide elastomer	PEBA	Pebax

4.3 Meet the Family

The material families described in this chapter represent a small sample of the commercially available thermoplastic materials. However, collectively these families represent the majority of materials that are currently in use around the world. While some of these materials are relatively new, many have been in use for decades, some for several generations. Used wisely, they have proven to be safe, reliable, and effective.

For most of us, it is unlikely that we will use a material from every different family, in every different category. Our day-to-day work often involves very specific expertise within a given price-to-performance category. It is very rare that one is asked to design an ultrahigh volume disposable medical diagnostic kit on a Monday, and then on Tuesday be asked to design a one-of-a-kind specialized test device that will go on the space station. These two design projects will have very different performance criteria, and the materials used will be very different (as will the processing technologies).

Let's go back to our conversation about looking at the world of thermoplastics from a user perspective, as someone tasked with selecting a single, specific material out of a possible list of tens of thousands of different materials. You could create a short list of performance requirements for the material based on property data in some specific areas. You could also input these requirements into a search engine, and then run some sorting protocols that use commercially available, published property data. If your requirements are correct, and if your search engine is smart enough—and your database is big enough—it will probably spit out some code names for some materials you might want to consider. However, depending on your search algorithms, and the quality of your input, the output could also be worthless (another classic example of *Garbage In, Garbage Out*).

An alternative method is to approach this task on a holistic level. This method involves a more human approach, based not just on the history of our species as users of materials, but also on an appreciation of the unique characteristics of thermoplastics, as well as an understanding of their behavior and performance as they are used in the world around us.

This approach is not for the faint of heart. It requires a willingness to set aside data sheets and property charts—at least for the moment—and focus instead on developing an intimate relationship with these materials. In order to compare and contrast and evaluate them, you have to get to know them, on almost a personal level. This is how humans have gone about making things, for thousands and thousands of years.

But how can we do this with plastics?

For starters, get familiar with the family of the material you are getting involved with. Get some samples of the materials in this family: test plaques, stock shape pieces (bars, bricks, rods, tubes), perhaps even some molded parts and experiment with them. Cut them with a saw (hand-saw or powered saw), and observe how they cut. Is the cut clean and uniform, or does the saw blade bounce and chatter? What sound do they make as you cut them?

Get a drill, and drill some holes of different sizes and at different angles. Try different drill types, and different drill speeds. Observe how the drill cuts into the material. Does the drill point wander, or does it bite into the material cleanly? What kind of chips form? Are they wavy and flaky, or do they break off in small pieces? Or does it form long spiral threads that wind up along the flutes of the drill bit? What happens when you pull on these long spiral pieces? Do they stretch and elongate (like that piece of silly putty we discussed), or do they simply break in half?

Also, experiment with how the material performs under abuse. Drop some material samples onto a hard, concrete surface. Throw them against a brick wall. Smash them with a rock, or a hammer, or an axe (use safety goggles and protective gear, of course). Take a pair of shears, and try cutting the samples. Do they cut cleanly? Or do they crack and split? Cut a bar halfway through, and then try and break the bar in half. How much effort did it take? Did it snap cleanly, or deform and elongate? Take some of these broken parts and look at the fractured surfaces under a microscope. Are the surfaces smooth and glossy? Or rough and textured? Are there peaks and valleys, or is it a flat plane? Then take some of these parts and stick them overnight in a refrigerator or freezer, and put some other parts in a warm place (like the dashboard of your car when it is parked in direct sun in the middle of the summer). Repeat the above experiments with these cold and hot parts.

These experimentation processes are NOT intended to be quantitative tests. There is no equipment to be calibrated, no test protocols to follow, and no measurements to be recorded. Rather, these experiments are intended to help you develop a familiarity with the material on a gut-level basis in order to provide a qualitative assessment of the age-old question, "How well do you know each other?" To quote an old Chinese proverb, *You don't really know someone until you have fought them.*

This knowledge represents the true essence of materials science. It is what has been missing in the use of thermoplastics, and the goal of this book is to fill that gap. In the following chapters, we will discuss some specific methods on how to properly select these thermoplastic materials, based on the specific needs and requirements of a given design application.

Further Reading

[1] E.A. Campo, Polymers in Industrial Applications, Hanser Gardner Publications, 2007.
[2] G.E. Dieter, Overview of the Materials Selection Process, ASM Handbook, Materials Selection and Design, vol. 20, 1997.
[3] H.-G. Elias, Macromolecules, Applications of Polymers, vol. 4, Wiley-VCH, 2009.
[4] E.R. Larson, Life Cycle Design Guidance Manual, U.S. Environmental Protection Agency, 1993 (contributor).
[5] E.R. Larson, Big, baad, and beautiful: advances in the use of engineering plastics in the office furniture industry, in: SPI Structural Plastics Conference, April 1990, November 1990. Reprinted in Modern Woodworking.
[6] E.R. Larson, Engineering Resins: Breaking the Rules, Plastics Engineering, May 1990 (editorial contribution).
[7] E.R. Larson, Finding New Parts for Plastics, Mechanical Engineering, August 1989 (editorial contribution).
[8] L.W. McKeen, Effect of Temperature and Other Factors on Plastics and Elastomers, second ed., Plastics Design Library, 2008.
[9] A. Michael, Materials Selection in Mechanical Design, third ed., Butterworth-Heinemann, Burlington, Massachusetts, 1999.

5 Material Selection Based on Performance

Material selection is a difficult task. Regardless of whether the material in question is wood, metal, stone, or plastic, selecting the proper material for a given application is a complex process. Before one even begins thinking about the materials, one must consider the requirements of the manufacturing processes involved, cost targets (and constraints), environmental concerns (in-use and post use), regulatory agency requirements, and often cultural and political considerations as well.

Then, as one begins to evaluate materials, one must consider chemical families, grades, versions, property data (and/or the lack thereof), testing and verification, agency approvals, sourcing and supply chain issues, and proper processing. Sadly, many engineers and designers short circuit the selection process by jumping immediately into property data, combing databases and material data sheets to find the highest value of one specific property in order to determine the best material for the application.

However, material selection is not about finding the "best" possible material for an application. Rather, it is about finding one or more suitable materials that—in combination with an effective design, proper processing, and eventual integration into a final system—result in a product that meets its intended use and satisfies (and hopefully delights) the needs of the end user. Far too often, in our quest to find the best material, we often forget that the real goal is to make the best possible product.

The ultimate goal of effective material selection is to optimize the performance of the product itself. While this may seem like a trivial statement, it is an important one.

5.1 What is Performance?

Performance is another one of those words that has a number of different meanings. In engineering, it is commonly used to describe the function of a system and how well it achieves its intended purpose.

When we talk about product performance, we are referring to an overall assessment of a product based on an evaluation of a number of measured parameters. For example, for an automobile we may measure acceleration, handling on the road, cornering, roominess of the interior, the sound levels while driving, and riding comfort. The performance criteria for a race car will be distinctly different than for a family sedan, or for a sports coupe. For sports equipment we may measure weight, stiffness, handling at high speed, vibration characteristics, the feel in our hands, as well as output at specific loading conditions (e.g., the launch angle and spin rate of a golf ball when struck by the clubhead of a driver at a specific head velocity). For a medical device we may measure the reliability and consistency of its operation under a wide variety of use scenarios, including mis-use (unintended or intentional).

One of the great challenges in design is in establishing the proper criteria for product performance. What parameters are going to be measured? What are the desired values for each parameter? How do each of these parameters contribute to the overall product performance?

Many companies have a formal process to develop these criteria. It usually begins with a list of product features based on marketing requirements, wish lists, desirables, and gotta-haves. It also often includes a list of *must not* requirements, such as the product must not cause injury when used in a certain way. Engineering requirements are then added to the list, addressing structure, durability, safety, etc. These typically also address the environmental conditions the product will be exposed to, and what the measured parameters must be under those conditions. Finally there are manufacturing requirements, including cost targets, and the desired levels of accuracy, precision, and overall quality.

As a result of this process, there will be a list of product specifications. This list should describe the criteria for product performance. Hopefully, every item on that list should clearly and specifically describe what is to be measured, how it is to be measured, and what the desired values are for that parameter. Done properly, a product that meets all of its specifications will have the desired product performance.

It is important to remember that performance is NOT an absolute measurement in and of itself. Rather it is a subjective evaluation based on a series of comparisons to an established benchmark. Benchmarks are an important tool, and not just in business analysis. Benchmarks help establish a set of expectations, a threshold for what is—and what is not—an acceptable performance level.

Several years ago, I needed to buy new tires for my car. I don't remember what car I was driving, but I was looking for a tire that would provide a nice ride and decent handling, and would also perform in all four seasons (since I live in southern California, "four seasons" may mean something different than in other parts of the world). Since I hate buying tires,

Figure 5.1 A set of new tires. ER_09/Shutterstock.com.

I wanted tires that would last, and while I wanted them to be reasonably priced, I was willing to pay a small premium for longer wearing tires.

I did a little research, and settled on a performance category titled, *Grand Touring All Season*. I started comparing brands and models, and read numerous reviews. The funny thing was, all of the reviews I read compared the performance of the tire under review to a tire made by Michelin, the MXV4. *The blah–blah tire offers a quieter ride than the MVX4, but it doesn't last as long. The yadda–yadda tire offers better handling than the MVX4, but does not perform as well in the rain.* At a certain point, I remember asking myself, "Why don't I just buy the MXV4 tires?" I realized the Michelin MVV4 was the benchmark for this category of tires. I bought a set the next day (Figure 5.1).

5.2 Predicting Performance

Predicting the overall performance of a new product is a challenging task. An analogy can be made from the world of sports. Every sport involves its own unique set of skills. In most sports there are coaches and trainers, whose jobs consist of developing and honing the skills of an athlete in order to achieve optimum performance. These coaches and trainers often rely on scouts and talent agents, whose job it is to find new athletes with promising athletic ability.

There are countless ways to evaluate specific aspects of athletic ability: strength, agility, hand–eye coordination, how high one can jump, how fast one can run, etc. Yet, at the end of the day, after all of the measurements are in, there is always a debate—which athlete will be the best performer?

The same can be said for product performance. Even with the best list of specifications, and the use of benchmarks, there are times when the product itself is less than the sum total of its parts. Why is this? I think there are several reasons. First and foremost, there is the issue of correlation.

5.2.1 Correlation

There are times when the correlation between a specific parameter and the product performance is clear. For instance, in an automobile of a specific size and weight, the more horsepower the engine has, the faster the car can accelerate, and the higher its top end speed (Figure 5.2).

There are other times when the correlation is not easy to determine. As an example, how does the relative stiffness of the membrane used to support the key pad on a laptop computer affect the overall product performance?

Furthermore, the issue of correlation extends not just to the parameter being measured, but to the properties of the materials that are used in the system. If we look at the intake manifold on the engine of a car, and instead of making it out of aluminum (perhaps using a die casting process)

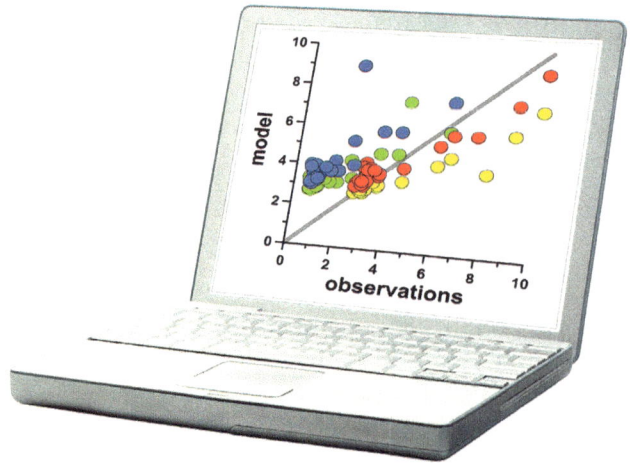

Figure 5.2 A correlation model. marekuliasz/Shutterstock.com.

we make it out of glass-reinforced nylon (using an injection molding process), will the engine have more horsepower? What other effects might this have on the engine? And on the overall performance of the car? Often these correlations are hard to determine. Also, changing one parameter may have unintended consequences in other areas of performance. For example, how would an engine with more horsepower affect the overall handling of the car?

5.2.2 Wrong Criteria

There are also times when we are simply measuring the wrong parameters. Going back to our sports example, let us look at the world of professional baseball. For decades, coaches and scouts and general managers relied on a traditional set of parameters to evaluate players. In the early 2000 season, the Oakland Athletics began to use a new method of evaluation, based on a completely different set of parameters. As described in Michael Lewis' book *Moneyball* [1] (later made into a movie), the use of these new parameters changed the sport.

Business managers and scouts in other sports have since adopted similar ideas, and are looking at and evaluating all kinds of performance data in all kinds of ways. We need to do the same kinds of things in the world of material selection. We need to make sure we are evaluating the right parameters, and we need to understand how those parameters correlate to actual product performance.

5.2.3 Disruptive Innovation

As much as we may enjoy working on new things, for most of us, our day-to-day job usually involves working on things that we are familiar with. In the engineering world, we may sometimes get involved in refining an established methodology, or in implementing a new and improved version of something, or in exploring a new technology. In most of these situations, there are examples of products in the real world that we can use to compare and contrast, either as benchmarks, or as a stretch goal of something to improve upon.

On rare occasions, we may be offered an opportunity to work on something completely new, perhaps even a product or technology that can change the world. These opportunities don't happen all that often, but when they do, they present a unique set of challenges. One of those challenges is in correlating material properties, evaluation parameters, and the performance of the new product—when nothing like it has ever been made before.

In other words, *how do you predict performance when there is no existing benchmark?* While I don't claim to have an answer to this question, you have to admit, it is a nice problem to have.

5.3 How Material Selection Affects Performance

As difficult as it is to correlate the effect of a specific parameter on overall product performance, it is even harder to determine the effects of a specific material that is used on a particular component in that product. Even if the evaluation criteria are perfect, the behavior of the materials used are sometimes so complex that it is impossible to determine which material property is making the difference. In most cases, it is not one specific property that makes the difference, but a combination of properties.

In the 1980s, there was a major effort among the major resin suppliers to seek out new applications by replacing parts and systems that had been traditionally made out of metal with parts and systems that were injection molded from engineering plastics (acetal, nylon, polycarbonate, polyester, etc.). These applications were in a wide range of industries, including automobiles, industrial equipment, household appliances, and office furniture [2].

One of the targeted applications in the office furniture industry was the classic five-legged chair base. Up until then, chair bases had been made out of pieces of tubular steel that were welded together, or out of a single large piece of die cast aluminum (or even zinc). In either case, there was not only the cost of the raw material, but the cost of fabrication, plus the cost of deburring and cleaning, followed by the cost of painting or plating or whatever secondary finish was required. Would it not be better to make a five-legged chair base out of an engineering plastic in a single part with a molded-in finish (Figure 5.3)?

DuPont did some investigation and was convinced that this could be done. In order to convince the experts in the office furniture industry, they went and had a mold made, and then fabricated parts in their own test lab using a very large injection molding machine. The initial prototypes were molded using a glass-reinforced polyester. This material was selected because it had exceptionally high tensile strength, as well as high stiffness. The resulting parts were stiff and strong, and met the basic performance requirements.

After some further evaluation, there was some doubt as to whether glass-reinforced polyester was the optimum material. So, as an exploration, DuPont molded some additional prototypes using a glass-reinforced

5: Material Selection Based on Performance

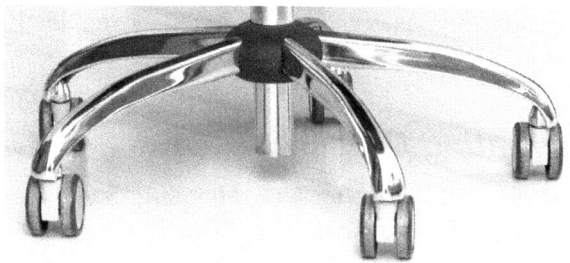

Figure 5.3 A traditional chair base made of metal. chaoss/Shutterstock.com.

nylon. These parts were molded in the exact same mold, with no design changes. However, these parts were distinctly different. They looked different, they felt different, they even sounded different. And they had different performance characteristics—almost all of which were better.

One of the interesting things was that when the parts were tested, a chair base molded in glass-reinforced nylon could actually withstand a higher ultimate load than a chair base molded in glass-reinforced polyester. How could this be? The stiffness and strength of glass-reinforced polyester is significantly higher. Processing was not an issue, as the parts were molded in a test lab under carefully controlled conditions. One of the theories postulated was that during loading, something was occurring on a localized level, where the stiffening ribs joined the main structural wall in each leg. At these junctures, there was a higher level of stress. In the glass-reinforced polyester, the local strain exceeded the maximum allowable strain, and a crack was initiated, which then propagated through the part, leading to structural failure. In the parts molded of glass-reinforced nylon, the material was able to yield, and the localized stresses were redistributed, allowing for a higher ultimate load. Another theory was that glass-reinforced nylon had better toughness than glass-reinforced polyester. (We will discuss the concept of material toughness in greater detail later in this chapter.)

Regardless of the exact technical phenomena, the bottom line was that a chair base molded from glass-reinforced nylon had better structural performance than a chair base molded from glass-reinforced polyester. Over time, and after extensive testing, it became obvious that chair bases molded from glass-reinforced nylon outperformed chair bases molded from glass-reinforced polyester (or any other thermoplastic material). Not only were they stronger, they could withstand impact better, they had a better surface finish, and they even sounded better when the chair was rolled across the floor. Today, some 30 years later, there are hundreds of thousands of chair bases made every year, and most of them are molded out of glass-reinforced nylon.

5.3.1 Evaluating Property Data

As challenging as it is to determine what specific material properties affect product performance, it can be even more challenging to quantify the effects of a change in the values of one or more properties. In other words, if you are evaluating material A and material B, and they have slightly different properties, what is the effect on performance by changing from material A to material B? This is an important question, but it can be arduous to evaluate.

In the initial phase of material selection, instead of focusing on the required value of a specific property, it is often easier to evaluate the types of material properties that are important. In other words, looking at mechanical properties (if there are structural requirements), or thermal properties (if there are temperature requirements), or properties related to toughness (if there are impact requirements). In this early phase, the evaluation is general and qualitative.

This initial phase may also involve a de-selection process. Quite often, there are some performance requirements that simply cannot be met by a large number of materials. Perhaps there is a structural requirement at high temperature, or an impact requirement at low temperature, or a requirement for long-term stability when exposed to a specific chemical (we will discuss some of these situations later in this chapter). These kinds of requirements can often be used to make a quick first cut to eliminate a number of material candidates. It is kind of like making the first cut when you are trying out for the varsity team, or auditioning for a role in a play. Many times this de-selection process is overlooked, but acknowledging it (and documenting it) can be an important tool later in the process, or in subsequent projects.

5.3.2 The Importance of Design

It is important to remember that material selection and design are interrelated. When you evaluate the performance of the end product, which is worse: a good design with the wrong material or a bad design with the right material? Neither is optimal. Furthermore, many design decisions one must make can affect the requirements of the material, and many material properties will affect your design decisions.

As an example, there are many times where the stiffness of a given part is an important criteria in the overall product performance. Often, those tasked with material selection will try to find the stiffest possible material. In the process, they may overlook the importance of other material properties, or the importance of good design.

The stiffness of a given design is easy to calculate—provided you have the basic dimensions in place (length, width, thickness, etc.), and you have an idea of the material you would like to use. There are a number of standard engineering equations one can use, based on classic beam and/or plate theory. As long as you understand the constraints, you can easily solve for any number of desired variables (deflection, stress, strain, etc.), simply by inputting some basic data.

In almost all of these equations, there are two important input variables. One of these is the elastic modulus of the material, E, as described in Chapter 3. This is then combined with the stiffness of the structure, which is determined by its moment of inertia, or I. E and I can be measured (or calculated) in any given direction, or in any mode of motion. In bending applications, the applicable modulus of elasticity is the flexural modulus, and the moment of inertia will depend on the shape of the structure. While the equations to determine it can be complex, I is almost always based on the cube of the thickness (Figure 5.4).

As an example, let us look at a cantilevered beam. It is one of the simplest structures. If we assume the beam has a load at the free end, the deflection at the tip is determined by the following equation:

$$y = \frac{PL^3}{3*E*I}$$

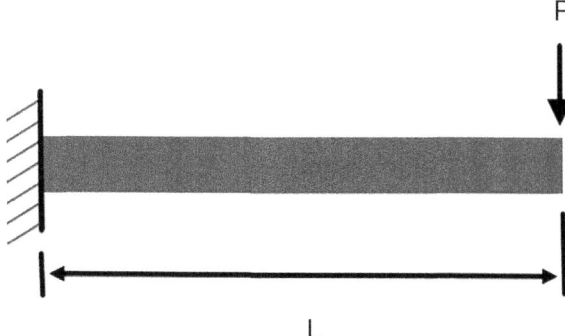

Figure 5.4 Cantilevered beam with end load.

where y is the deflection, L is the length, of the beam, and P is the force applied at the tip. For a beam with rectangular cross section, the cross sectional moment of inertia I is calculated as follows:

$$I = \frac{b*h^3}{12}$$

where b is the width and h is the height (or the thickness). From the deflection equation, the deflection is inversely proportional to the product of E times I. Both E and I are input variables and are a result of the stiffness of the material times the stiffness of the structure. In essence, this EI product is a design variable. Some refer to this variable as "the stiffness factor," but I like to call it the Old McDonald factor (from the children's song *Old McDonald had a farm, E, I, E, I, O!*).

What is interesting to note is that if you double the value of E—that is, if you select a material that is twice as stiff—you will cut the deflection in half. However, if you used the same value of E, and instead increased the thickness by 25%, the value of I would almost double, which would also cut the deflection in half (actually it would reduce it by 48.8%, since 1.25 cubed = 1.953, and 1/1.953 = 0.512).

Selecting materials for stiffness—based solely on the published value of their flexural modulus—is often counterproductive. It overlooks the stiffness contribution of the structure itself and neglects to account for minor changes in the design—many of which can have major effects on the overall structure.

While the relationship between stiffness and thickness is straightforward, there are many other design–material relationships, some of which are quite complicated. These relationships need to be addressed during the material selection process. (We will explore some of these relationships in greater detail later in this book.)

5.3.3 The Importance of Processing

Just as material selection and design are interrelated, the processing of thermoplastic materials also affects product performance. One of the primary reasons is that processing affects the properties of the material.

First and foremost, you need to remember that published material property data are generated from test samples. These samples were made from prime, 100% virgin resin, and were carefully prepared under controlled conditions, and then tested under controlled conditions. They were not molded by a production molder who was trying to optimize the molding process for maximum cycle time efficiency and lowest production cost. Furthermore, the test lab

5: Material Selection Based on Performance

was not tasked with minimizing warpage or maintaining specific tolerances or achieving a high gloss Class A surface with no visible cosmetic defects. And the test lab was probably not all that concerned about the size and location of the gate, or of the ejector pins, or the amount of visible flash at the parting line.

In a production environment, most plastic processors are tasked with delivering plastic parts or subassemblies which contain plastic parts. They usually have very specific requirements that they are responsible for (usually dimensional requirements, but also on occasion other performance criteria). However, very rarely are they held directly responsible for the material properties of the material itself in the final molded parts. It is often impossible to measure those properties, to say nothing of correlating the actual in-use material properties in the molded part to the part performance (just as it is in the design phase).

One of the main reasons that material properties are affected is due to changes in the molecular weight distribution of the polymer chains in the molded resin. This is especially true in processes like extrusion and injection molding where the resin is taken to a molten state by the application of shear and high pressure. Even with processing methods where the material does not undergo high rates of shear—(such as thermoforming, rotational molding, pressure molding, or ultrasonic welding)—the properties of the material can be affected. What properties are affected, and how they are affected, will depend on the materials involved, and what production processes are involved. The important thing to remember:

> Every plastic manufacturing process can have an effect on material properties.

We will explore how to account for this phenomenon in the material selection process in a later section.

5.3.4 Property Data—A Final Caveat

Time and time again I encounter someone who asks me a variation of the following:

> I am looking for a material where property ABC has a minimum value of 123 units. What material do you recommend?

As a general rule, I try not to get involved in these kinds of questions. It is a sign that the questioner is not experienced with plastic material selection. As a consultant, and someone engaged in new business development, you might ask, *How could you pass up a new client opportunity?* The truth of the matter is, working with clients like this is often more of a bother than it is worth. And sometimes the questions they ask make you shake your head.

The fact of the matter is, you cannot select a material—any material—based solely on the measured value of one single material property. You have to evaluate a number of properties, and correlate them to the desired performance of the end product.

Recently, I came across a post in an online forum where someone was looking for an injection molding resin with a heat deflection temperature (HDT) of 450 °F or higher. At first glance, one might think this is a challenging question being posed by a very sophisticated user of materials. However, it is really nothing more than a variation of the ABC-123 question.

However, one of the things I have learned in my career as a consultant is that there is a distinct difference between saying, "That's a stupid idea" and asking the question, "Can you tell me a little bit more about the decision process you used when you selected this particular material?" In this case, I did not have the opportunity to ask that question, but in looking behind his focus on HDT, it was obvious that they were looking for a specialty material, one that had exceptional performance at some very high temperatures (most likely excellent strength, stiffness, etc.). However, he made no mention of any other performance requirements (chemical, environmental, etc.), nor did he describe the application in any detail. So how could anyone possibly recommend the absolute best material?

But rather than lecture him on that, I simply advised him to not focus on HDT, and suggested he carefully evaluate other performance requirements. I encouraged him to look at LCP, polyimides (PI, PEI, PAI), polysulfones, the ketone family (PEK, PEEK, PEKK, etc.), PPS, and perhaps even high-temperature nylons (HTN, PPA, etc.).

Ironically, many of these specialty materials are used in the aerospace industry. However, while evaluating their suitability for a given application may take a little bit of effort, it is not exactly rocket science [3].

5.4 Environmental Effects

In any material selection process, one must consider how the environment that a product is used in will affect its performance. One aspect of this is in the measurement of specific performance criteria, such as the road handling of a car using a specific tire on a wet road. On a material selection level, we are not only concerned with how the specific tire performs, but also with how the environment affects the material used in the tire. There are a number of environmental phenomena that can affect materials. These effects can be loosely grouped into two main categories: those that are reversible and those that are not.

Reversible changes in materials happen all the time, in all types of materials. As an example, almost all materials expand with the application of heat and contract when they are exposed to cold. Also, most materials become more flexible at high temperature and get stiffer when they are cold. Some materials may soften when they get wet, but will return to their original hardness when they dry out. These types of changes are common, and under most situations they are fully reversible as long as the material has not gone through a permanent phase change (such as cement turning into concrete). While these kinds of reversible changes need to be accounted for in the selection of the material (and in the design), they are normally not a big deal.

What is a big deal is when exposure to the environment causes irreversible changes in the material itself. These changes include chemical reactions, structural changes in the polymer matrix, degradation of the polymer, and sometimes even a complete depolymerization of the polymer molecules (a breakdown of the polymer chain into its base monomers).

The environmental factors which cause these changes can be grouped into four main areas: temperature, chemicals, radiation, and time. Exposure to any of them—and all of them—can wreak havoc on the material properties of thermoplastics. I like to call them *The Four Horsemen of the Plastic Apocalypse* (Figure 5.5).

Figure 5.5 The four horsemen of the plastic apocalypse.

5.4.1 Temperature

As described in an earlier chapter, all thermoplastics soften (and/or melt) at high temperature. However, even at temperatures much lower than T_g or T_m, long-term exposure to heat will have a detrimental effect on a thermoplastic material. The primary reason is that this exposure to heat causes a breakdown of the polymer chains, resulting in a lower molecular weight distribution and a loss of properties. The most common losses are in elasticity and toughness, but other properties are affected as well. The temperature at which this degradation begins to occur will vary, depending on the chemical family of the polymer, as does the exact chemical mechanism involved (oxidation, depolymerization, etc.). Occasionally, this degradation can be reduced through the use of additives known as heat stabilizers. It still occurs, but at a higher temperature and at a lower rate.

One commonly referenced material property is known as the HDT. This is a standard test where a specimen of a material is subjected to a defined load and then slowly heated while measuring the deflection. As the material gets warmer, it becomes less stiff and the deflection will increase. Once a defined amount of deflection is achieved, the test is complete and the temperature is recorded. This temperature is the HDT for that material. Occasionally, the HDT will be measured using different loads (most data sheets will reference the load along with the measured HDT).

The test itself is simple to conduct, however, the test is merely an assessment of material stiffness at elevated temperature. It is NOT an assessment of the actual service temperature of that material, nor does it make any prediction of polymer degradation. Also, since the test is short term in nature, it should not be used to evaluate long-term performance.

A more useful piece of data for material selection is the Continuous Service Temperature (sometimes also referred to simply as Service Temperature, and also known as the Relative Thermal Index). It is the highest temperature at which a material can function for an extended period of time without failing. Unfortunately, the Service Temperature is often difficult to determine. What amount of time is "an extended period?" And what functions need to occur without failing? There are some defined tests which can be used to quantify service temperature, based on electrical properties, mechanical properties, etc. And while it is easy to detect major differences in service temperatures between materials, it is often only through extensive testing that one can quantify the long-term performance of a given material at a given temperature.

At the other end of the temperature spectrum, thermoplastics are also affected by extreme cold. Most of this effect is seen in brittleness, in that there is a loss (sometimes a complete loss) of ductility, and even low stresses will cause brittle fracture. While there may also be some polymer degradation at extremely low temperatures, these phenomena are rarely studied.

5.4.2 Chemicals

Like most materials, thermoplastics are also susceptible to chemical attack. Normally, when we think of chemicals, we think of acids and bases, alcohols, gasoline and other fuels, solvents (paint and lacquer thinner, acetone, toluene, etc.), and detergents and cleaning solutions. But there also chemicals in fats, oils, greases, lubricants, pesticides, and disinfectants. And then there is salt, not just the standard sodium chloride in sea water or table salt, but an entire category of chemical compounds, some of which are found in nature, others which are synthesized. Then there are airborne chemicals, gases and vapors and fumes, oh my!

The manner in which a thermoplastic material is affected by exposure to a given chemical depends on a number of variables. First and foremost is whether the thermoplastic reacts with that chemical. It may be completely impervious to that chemical, no matter what. Or it may be unaffected at low temperature, but affected by exposure at high temperature. Then there is the relative concentration of the chemical, whether the exposure is constant or intermittent, and the duration of the exposure. Finally, there is the chemical mechanism involved. Is the chemical acting as a plasticizer, and if so, is it a reversible action, or permanent? Is the chemical causing an oxidation reaction, polymer degradation, or simply a discoloration of the surface, etc.?

While some of these questions may involve some detailed testing and analysis, most resin suppliers will publish some test data of the affect of chemical exposure on some basic material properties, such as stiffness, tensile strength, etc. They will also publish guidelines on whether a given material is suitable for use with, not suitable for use with, or slightly affected by various common chemicals.

One chemical that is often overlooked is H_2O, water. Most of us think of water as an inert material, but for some materials, such as raw iron, exposure to water causes an immediate chemical reaction. Fortunately, most thermoplastics do not chemically react with water. But there are some thermoplastics, such as nylon, which absorb water. This absorption process, which is fully reversible, causes the material to swell, and also acts as a plasticizer, making the material tougher, more flexible, and more ductile although it also reduces its strength.

However, if you can combine exposure to water with exposure to high heat, many thermoplastics will decompose as the bonds in the polymer chains are broken down. This process is known as hydrolysis, and literally means water splitting (from the Greek words *hydro-*, meaning "water," and *lysis*, meaning "separation"). The temperature at which this occurs depends on the thermoplastic.

Water can also act as a solvent for other chemicals. In those situations the exposure to water is not the issue, it is the chemical(s) contained in the water. Regardless of whether the water is used for irrigation or potable use (or other), it is important to know its source. Understanding the source— be it a well, a river, a stream, a lake, a dam, or even the ocean—will provide insight into what chemicals and minerals it may contain. Even tap water can contain chemicals, as municipal water treatment centers in various parts of the world frequently treat the water to remove pathogens. So if you are selecting a material that will be in contact with water, you need to be aware of what chemicals could be in that water.

Another aspect of chemical reactivity is flammability. Flammability is an assessment of how easily a material will ignite and burn. While combustion (the act of burning) is a complex phenomenon, it is basically a series of chemical reactions. These reactions involve polymer decomposition, gas generation, oxidation, and more. While I would not categorize the introduction of an open flame as a chemical exposure, I think you would agree that the act of combustion would have a detrimental effect on the material properties.

5.4.3 Radiation

Another environmental factor which affects thermoplastics is radiation. Most people think of the term radiation as it pertains to radioactivity, which describes a material which emits particles and energy as part of nuclear decay. But radiation is actually a much broader term, and describes the process by which electromagnetic waves travel through space.

Electromagnetic waves are a form of energy that is composed of an electrical field and a magnetic field. These waves can have a wavelength of as small as 1 pm (10^{-12} m) to as large as 100 Mm (10^6 m, or 1000 km). This range of wavelengths, commonly known as the electromagnetic spectrum, begins with gamma rays (at 1 pm), and includes X-rays, ultraviolet UV light, visible light, infrared, microwaves, and radio waves (Figure 5.6).

The amount of energy carried by these waves decreases as the wavelength increases. Gamma rays carry the most energy, followed by X-rays, then UV light. In physics, EM waves are collectively described as "light" waves, although the term "light" typically is used to describe visible light which are electromagnetic waves with wavelengths between roughly 390 and 750 nm (Figure 5.7).

5: Material Selection Based on Performance 161

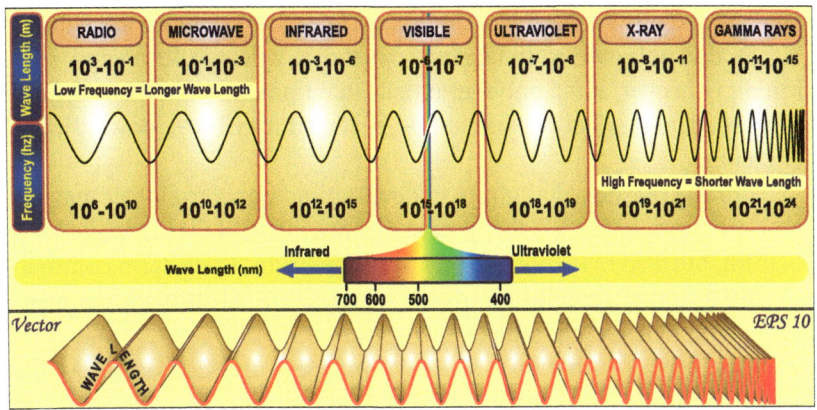

Figure 5.6 Electromagnetic spectrum. Fouad A. Saad/Shutterstock.com.

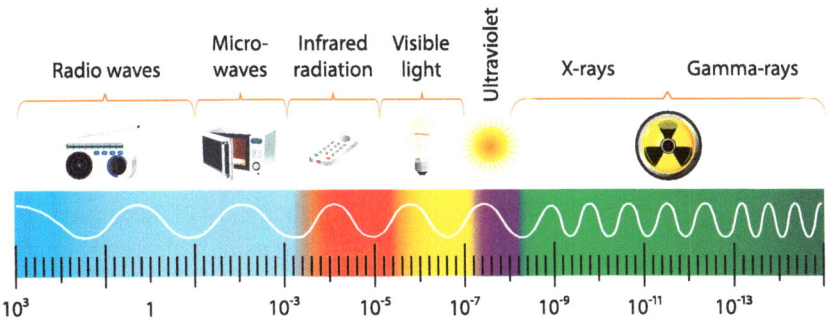

Figure 5.7 The electromagnetic spectrum, shown in order of decreasing wavelength and increasing frequency. Designua/Shutterstock.com.

In thermoplastic material selection, we are sometimes concerned with whether a given thermoplastic—and the additives it contains—will block a given frequency of EM waves, or transmits them without loss. For example, in optical applications, we typically want all light in the visible spectrum to be transmitted, without concern for other wavelengths. Or, in the case of sunglasses, we may want to block a certain amount of visible light or wavelengths in the UV range. Or, in an electronic shielding application, we may want to block transmission of EM waves in a certain band of the radiofrequency (RF) spectrum.

However, we also need to account for the effects of any EM waves on the polymer itself. Basically, we are putting energy into the polymer matrix, especially at the lower end of the spectrum (gamma rays through UV). If the polymer is transparent to those waves, the energy passes through. However, if the polymer blocks that transmission, the energy will be absorbed, and either converted into heat or it may break down the polymer chains.

One of the reasons sunlight is so devastating to materials (all materials, not just thermoplastics) is that it contains not just EM waves in the visible spectrum, but also in the infrared and UV spectrum. Long-term continuous exposure to direct sunlight means the material will absorb a lot of energy, usually with detrimental effects.

Another type of radiation is an electromagnetic pulse, or EMP. An EMP is a short burst of very high intensity and can be caused by a number of factors, including lightning strikes, electrostatic discharge, electrical power surges, and solar flares. There are also man-made EMP events, such as nuclear explosions, and the discharge of high-energy weapons (which may cause a non-nuclear electromagnetic pulse). The effects of this type of radiation on thermoplastics have not been well documented.

5.4.4 Time

The final environmental factor, and in some ways the most critical, is time. Time, in combination with one or more other environmental effects, will almost always result in polymer degradation. In fact, most of the test data that is used to evaluate environmental effects is created using time as a variable.

For instance, heat aging tests, which are used to evaluate the effect of long-term exposure to elevated temperatures, can be used to show the change in a given property value, say tensile strength, as a function of time. The graphs show sample data from different versions of nylon (Figure 5.8).

In a similar manner, weatherability tests are often used to assess the long-term effects of exposure to an outdoor environment. These tests typically address a combination of temperature, chemical, and radiation (primarily UV) effects, measured over the course of days, weeks, months, or years. These tests may include a variety of factors: for instance, an Arizona weathering test typically addresses high heat and high UV in a dry environment, while a Florida weathering test addresses high humidity and high UV in a subtropical environment, sometimes with the added effect of salt spray. While these tests are often conducted on an accelerated time scale, the intent is to predict long-term performance over months and years of exposure.

What is unknown is whether time in and of itself causes polymer degradation. In other words, do the polymer molecules in a thermoplastic

5: Material Selection Based on Performance

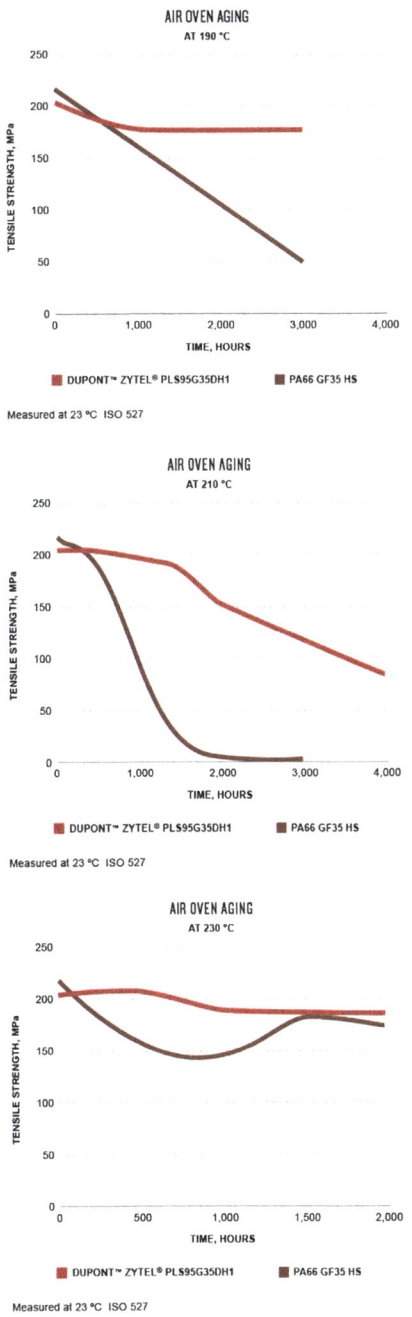

Figure 5.8 Effects of oven aging on nylon 6/6.

material degrade as a function of time? (Without exposure to heat, chemicals, or radiation?)

To evaluate that, we would need to make a special time capsule. Let us say we take four materials—a piece of wood, a piece of stone, a piece of steel, and a piece of plastic—and we place them in glass container. We seal the container, pump out the air inside, and as a precaution, re-fill it with nitrogen. Then we bury this container inside a deep cave, where the ambient temperature is a constant 73 °F, and there is no sunlight, and no radiation of any kind. Then we come back in 10,000 years.

So here we have four materials—two natural and two synthetic, two made of minerals and two made of polymers—that have been exposed to nothing but an inert gas, and time. What would we find inside that container? I would surmise that the piece of steel would be unchanged, chemically identical to time zero. I would assume the piece of stone would also be unchanged, although perhaps some of the chemical bonds that bond the various minerals together might have broken down. Perhaps we would see some dust particles as a result. However, I would expect the piece of wood might have some changes. Wood is made from the fibrous tissue of a tree, and consists of cellulose and lignin, two naturally occurring polymers. For some reason, I would expect the tissue in the wood to break down over time, even in the absence of any outside agent. So while there might be some structure to the piece of wood, I would expect to see a lot of sawdust. But what about the piece of plastic? Would it be completely intact, chemically identical to the original? Or would it decompose into a fine powder? This is an interesting question, and I don't know the answer. I also think the answer has implications to the future of our planet.

5.5 Key Mechanical Properties

When it comes to thermoplastic material selection, there are three primary mechanical behaviors that should always be considered: strength, stiffness, and toughness. Knowing the properties of a given material in these three areas will provide a fundamental understanding of the structural performance of the end product.

5.5.1 Strength

The strength of a material is its ability to withstand an applied force without failure.

5: Material Selection Based on Performance

At first glance, selecting materials based on strength seems like a straightforward task. You look at the structure involved, determine the loads, calculate the stresses, and then pick a material with sufficient strength. Property data on tensile strength is readily available for almost every material known to man, often at different temperatures, and under different environmental conditions.

The problem with this approach is that most products are three-dimensional objects, and are subjected to forces in all kinds of different directions. Unless you are dealing with rope or fishing line (in which case tensile strength is the critical property), you need to evaluate the strength requirements under a number of different loading conditions (Figure 5.9).

There are a number of methods of doing this. A common means is structural analysis using the finite element method. This type of analysis is well suited for identifying areas of high stress. A key question to then ask, *What is the primary stress state of the material in these high stress areas (tension, compression, or shear)?*

Once the stress state is understood, one can then look at the appropriate property data, whether it is tensile strength, compressive strength, shear strength, or bending/flexural strength. Also, one needs to consider which is more important, the yield strength (the point at which yielding occurs) or the ultimate strength (the strength at the point of failure).

Another important strength property which is especially important for injection molding applications is knit line strength. A knit line, also called a weld line, occurs when two flow fronts in a mold come together. This normally occurs when there are holes in the molded part and the flow front separates to flow around the steel that forms the hole, and then reconnects and knits on the opposite side. The strength across a knit line is almost always less than in the base material. In addition, if the material has any

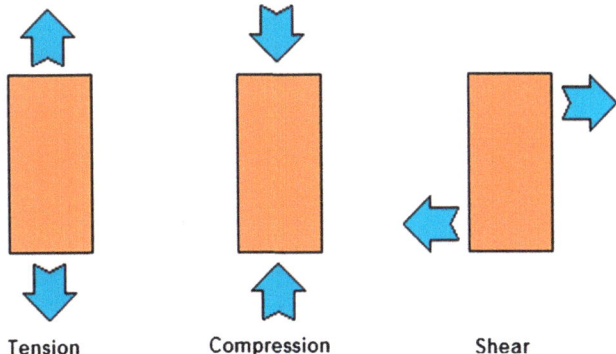

Figure 5.9 Types of loads.

reinforcement (glass fiber, mineral, etc.), the reinforcing agent will NOT cross the knit line, and the knit line strength will be dependent solely on the strength of the base material.

Unfortunately, there is very little published property data for knit line strength. Knit line strength is also heavily dependent on processing conditions.

Regardless of the type of strength, it is important to remember that strength is always evaluated on a force per cross sectional area basis. Sometimes, in an effort to find the highest strength material, we may forget that we could increase the ultimate load of the structure simply by increasing the cross sectional area, perhaps by increasing the thickness. Conversely, we may be limited on the size of the structure in certain cross sections, and must therefore select a material based on the loads in those specific areas.

5.5.2 Stiffness

Stiffness refers to the ability of a material to withstand an applied force without deformation. Typically, we think of stiffness as how well an object resists bending. In this scenario, the stiffness of the object is dependent not only on the flexural modulus of the material, but the moment of inertia of the structure itself.

Thermoplastic materials are often used in applications where resistance to bending is an important performance parameter. Furthermore, determining the flexural modulus of a material is relatively easy, and there is published property data for most thermoplastic materials. For most applications, selecting a material with the appropriate bending stiffness is a simple task. Again, it is important to remember that the stiffness of an object is also highly dependent on design.

However, there are situations where flexural modulus data is simply not available. This is especially true with fiber-reinforced materials, in which the flexural modulus of the material varies in different directions, depending on the orientation of the fibers. Often times the fibers will be oriented in the direction of flow, resulting in a much higher modulus in that direction than when measured across the flow. Property data for reinforced materials is typically prepared from molded test specimens, which have fiber alignment in the same orientation as the length of the test specimen. To measure the modulus across the flow, the test specimens need to be machined from a larger part (usually a plate or sheet) and cut so that the fiber orientation is perpendicular to the length of the test specimen. This adds time, and cost, and requires additional testing. Quite often, this testing is not done, and property data for cross flow flexural modulus is simply not available.

Furthermore, it is important to remember that the intrinsic stiffness of a material is much more than just the flexural modulus. As discussed earlier in Chapter 3, the stiffness of a material is dependent on the relationship between stress (due to the applied loads) and strain (the amount of deformation that results). For any given stress load, materials that have low strain are described as being stiff, and those with high strain are described as being flexible. This is true regardless of whether we are evaluating bending, tension, compression, or shear.

So, in order to evaluate materials based upon stiffness, we often have to look beyond flexural modulus, and evaluate tensile modulus, compressive modulus, and shear modulus as well. Sometimes, it is helpful to look at the actual stress–strain curves. As in the example of the plastic chair base mentioned earlier, we may find that in our quest for structural performance we have made an unrealistic requirement for high stiffness, when what we really should be looking for is an overall balance of strength, stiffness, and toughness.

5.5.3 Toughness

The toughness of a material is its ability to withstand sudden impact (Figure 5.10a).

Unlike strength and stiffness, which are evaluated using measurements based on force, toughness is evaluated using measurements based on energy, which means that not only are forces involved, but units of length and time as well. The toughness of a material is basically the amount of energy it can absorb without breaking, either through a brittle or ductile failure.

Just as there multiple ways to evaluate strength and stiffness, there are multiple ways to evaluate toughness, and there are a number of standard tests that are used to quantify the toughness of thermoplastic materials. Some of these tests are simple, and can be easily performed on a bench top with minimal equipment. Some tests are quite sophisticated and involved advanced equipment and extensive instrumentation. Regardless of what

Figure 5.10a Sudden impact. Alex Mit/Shutterstock.com.

tests are used, it is important to understand exactly what aspect of toughness is being measured.

First, what is the stress state of the material as it is being impacted? It is a pure tensile impact? A shear impact? Or does it involve a combination of stress states? Second, what is the mode of failure of the test specimen? In many impact tests, the mode of failure is a brittle fracture. This is important because brittle fracture is often due to cracks.

Fracture mechanics is a field of mechanical engineering that studies how cracks form and propagate in various materials. It is a complex and important field of engineering. Understanding how cracks form and propagate in steel is critical for the construction of bridges and buildings. For airplanes, the same holds true for aluminum. The formation and propagation of cracks in thermoplastics is a science in and of itself.

Regardless of the material, crack initiation and propagation is an important aspect of toughness. In some materials, such as window glass, cracks propagate easily. Indeed, the standard means of cutting window glass is to lightly score the surface, creating a crack or notch. Then, with a slight tap, the crack propagates, and the window pane breaks off. The term *notch sensitivity* is used to describe how a material responds to this type of crack propagation. Some thermoplastic materials have high notch sensitivity (although perhaps not as high as window glass). Other thermoplastic materials have low notch sensitivity (Figure 5.10b).

Another aspect of toughness is tear resistance. While we often think of this in relation to materials in a film or sheet form, it also applies to thick solids. The typical chew toy that you might give your pet is most likely from a material with high tear resistance (Figure 5.11).

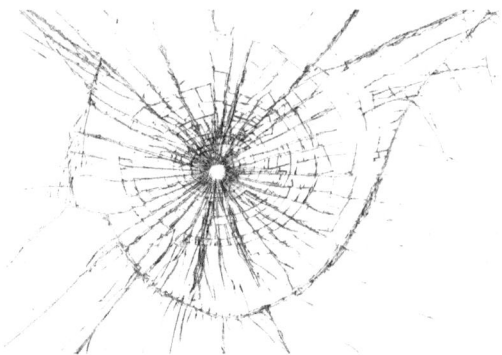

Figure 5.10b Window glass is notch sensitive. JoLin/Shutterstock.com.

Figure 5.11 Chew toys require good tear resistance. Mila Atkovska/Shutterstock.com.

Selecting a thermoplastic material based on toughness is a complex task. I would describe it as the single most difficult task in the field of plastics engineering. There are many reasons for this.

Impact analysis involves a huge number of variables: the stress states of the material(s) being impacted (tension, compression, shear, etc.); the rate of loading (i.e., the speed at which the impact occurs); the overall energy of the impact (involving not only the speed of impact but the masses of the components in the systems, including both the system that is delivering the impact and the system that is absorbing the impact); along with dozens of other variables. Identifying each of these variables can be a difficult task, to say nothing of what is involved in isolating and measuring the values of each variable in the system.

We also need to understand that impact analysis is an imprecise science. Even if we focus exclusively on the behavior of the material(s) under impact, there are dozens—if not hundreds—of assumptions on the behavior of the components in the system prior to and during impact. The validity and accuracy of these assumptions is often a topic of heated debate, even among experienced professionals. The behavior of the material(s) under impact will also be affected by temperature, and/or other environmental factors (exposure to chemicals, radiation, etc.). Quite often, the best assessment of impact is real-life testing of the actual system under the appropriate end-use conditions.

As a result, quantifying the toughness of a given material in a given application is a highly subjective assessment. Sometimes the toughness of a material is evaluated based on structural failure as the result of sudden impact, by looking at the loads and structural deformations when a device or test specimen completely fails. There are also situations where a device or system is subject to repeated loads, with forces lower than what would cause failure in a single impact. (Think of repeatedly hitting something with a hammer until it eventually breaks.) This type of toughness, commonly described as impact fatigue, is evaluated based upon the damage progression as the specimen undergoes repeated impact. In this case we need to account for not only all of the previously described variables, but the frequency of the loading (how often the impacts occur), and how many impacts occur. At an extreme level, one could say that vibration is a type of repeated impact. There are entire fields of engineering dedicated to vibration.

Regardless of the science involved, the fact remains that toughness is an important issue for many applications where thermoplastic materials are being considered. Quite often, the toughness of a material is the most important factor in product performance.

5.6 Measuring Toughness

There are a number of standardized tests that are used to quantify various aspects of material toughness. The following section discusses some of the more commonly used tests.

5.6.1 Izod Test

The Izod test is typically a bench top test. (The test is named for its inventor Edwin Gilbert Izod, and has nothing to do with a famous clothing maker.) In this test a small test specimen is clamped in a vise. This specimen has a V-shaped notch in it, facing forward. A pendulum arm is then raised to a certain height and released. The arm swings into the specimen, breaking it, and then continues swinging to the opposite side. A needle on a dial measures how high upward the arm swings. The difference in height between the starting position and the end position is used to calculate the difference in energy, which is the energy that was absorbed by the test specimen as it broke. The speed of the impact depends on the length of the arm, and height from which it is dropped.

The Izod test is a simple test to conduct, and Izod test data is readily available. It is important to remember that this test is a measurement of impact in almost a pure shear loading condition, and the specimen has a substantial notch. While the resulting data may be useful for material comparison, it represents an unusual end-use loading condition (Figure 5.12).

5: Material Selection Based on Performance

Figure 5.12 Schematic diagram of a typical pendulum test.

5.6.2 Un-Notched Izod Test

This test utilizes the same equipment and procedure as in the Izod test, except the test specimen has no notch. While the load case is still a pure shear loading condition, the fracture of the test specimen is independent of any notch. This test allows one to quickly compare impact data for a given material in a notched versus un-notched state. This comparison can provide insight into the notch sensitivity of the material.

5.6.3 Charpy Test

The Charpy test is similar to the Izod test. It is named after Georges Charpy, a French engineer and scientist who developed and standardized the test methods in the early 1900s. It is also a pendulum type test, but in this test the specimen is clamped sideways, by securing it at each end. The specimen could have a V-shaped or U-shaped notch, and the notch faces away from the pendulum. The test device itself could a bench top size, or a larger, floor size model.

The Charpy test is also simple to conduct. Charpy data is not as widely available as Izod data, but is usually easy to obtain. The loading is in pure shear. The data is useful for material comparison, and for evaluating the notch sensitivity (Figure 5.13).

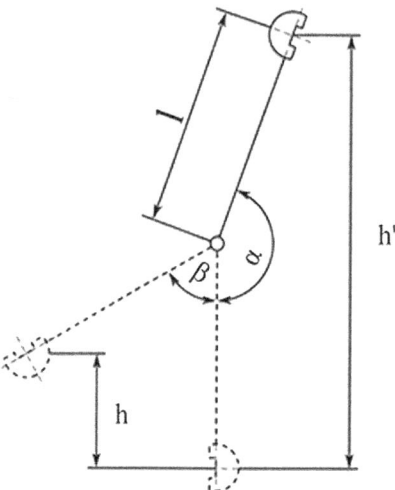

Figure 5.13 A Charpy-type impact test machine.

5.6.4 Gardner Impact Testing

Gardner impact testing refers to a type of impact testing that involves a weight that is dropped onto fixed object. Generically known as falling weight testing, the name Gardner comes from Paul N. Gardner, founder of the Paul N. Gardner Company.

The typical Gardner tester is a floor-based test, and uses a falling dart rather a pendulum. The "dart" is typically a weight with a rounded nose. The test specimen is a plaque, and rests on a plate with a hole in the center of a prescribed diameter. The dart is lifted to a specific height, and then dropped on the specimen. If the specimen breaks, the test result is recorded as Fail, if the specimen is intact, the test result is recorded as Pass. A typical test scenario involves a dart of a specific weight, a defined drop height, and the number of drops before the specimen breaks. The amount of energy is calculated based on the weight and height.

Gardner impact tests, and other similar falling weight tests, are simple to conduct. The equipment is relatively inexpensive, and while the data can sometimes be hard to compare (due to different weights, drop heights, and/or the number of Pass/Fail cycles), it does represent real-life conditions. Products get hit by falling objects all the time (or they are falling objects themselves), and this test can provide useful insight on impact under those conditions.

One minor caveat, since the weight is free falling, it is accelerating under gravity. So, the impact velocity of the weight will increase as the height increases. Since plastics are rate sensitive, this will affect the test results. One could have very different impact performance between the impact of a heavy weight falling from a low height and a lighter weight falling from a higher height, even though the kinetic energy is the same. This might be an important issue in some applications.

5.6.5 Instrumented Impact Tests

As the name implies, instrumented testing involves the use of equipment which can precisely record the variables involved in the test. It is a general term, and can be used to describe a wide variety of tests. For instance, one could attach some sensors and gauges to an Izod tester, connect them to a laptop computer, and then record the data as the pendulum is released and impacts the test specimen. Technically, this would be an instrumented impact test. However, the phrase *Instrumented Impact Testing* is normally used to describe testing where one cannot only measure and record the test variables, but can also have precise control over the actual test parameters, such as the rate of loading and the velocity of impact.

While instrumented impact testing can be quite simple (as in the instrumented Izod test described earlier), the real value of instrumented testing lies in the ability to isolate and control the various test parameters and to precisely record the output data. When applied properly, instrumented testing can provide valuable data on almost aspect of material toughness. The primary disadvantage of instrumented testing is that the equipment can be quite complex (and expensive).

5.6.6 High-Speed Tensile Tests

High-speed tensile tests are a form of instrumented impact testing. (They are the high-tech version of playing with silly putty.) They involve the use of standard tensile test specimens and standard tensile test equipment. However, traditional tensile tests are typically conducted at a very low rate of strain, typically ranging from fractions of inches per minute to about 20 in/min. (Even 20 in/min is a very slow rate of movement. It is less than 0.02 miles/h, or approximately 0.03 km/h. Compared to these rates, most turtles are moving at light speed.)

In high-speed tensile tests, the rate of loading is substantially increased, often times to a rate of loading that is hundreds—or even thousands—of times faster than in traditional tensile tests. While these rates of loading are quite common in the real world, in the traditional world of material science these rates of loading are unheard of.

The advantage of this type of testing is that it utilizes existing equipment and can be easily standardized. It can also provide a comparison of to real-life end-use conditions. Also, while comprehensive test data is not available for all materials, it can easily be generated on an as-needed basis. The disadvantage of this type of testing is that it is new and different. There is very little material property data available, and even with the data that exists, the correlation between measured data and actual end-use performance is still uncertain.

5.6.7 Projectile Testing

Projectile testing involves taking an object, and then, through the application of force by the test device, sending that object through space to impact against a test specimen. The object being projected could be a weight, a steel ball, a brick, a piece of organic matter (such as a head of lettuce or a frozen bird), and the force could be applied by compressed air, hydraulics, or other mechanical means. The test specimen could be a standard test specimen, a material sample, or a prototype or production part.

While projectile testing is common, the tests and equipment are not standardized. In general, resin suppliers to do not provide any test data for projectile testing. Instead, projectile testing is usually done with customized equipment with a specific intent, such as evaluating the impact resistance of automobile fascia when impacted by road debris, or of aircraft windshields when subjected to bird strikes.

The advantage of projectile testing is that the test can be developed to mimic the actual end-use conditions and materials can be evaluated in that context. The disadvantage is that the equipment, and the actual tests, are not standardized, and it may not be possible to correlate the results of these tests with other types of toughness tests (Figure 5.14).

5.6.8 Drop Testing

Drop testing is another type of impact testing involving a falling object. However, unlike other impact tests, the falling object is the actual test specimen. The test specimen is typically a fully assembled product, but could also be a subassembly, or a single part.

Drop testing can be conducted at a variety of heights, and the test specimens can be dropped onto different surfaces (concrete, hard packed dirt, carpeted floors, etc.). Just as in falling weight tests, the impact energy will be dependent on the height of the drop and the mass of the object. The

5: Material Selection Based on Performance

Figure 5.14 A drop test? Or a Gardener test gone wrong?

testing can also be used to assess the impact performance of the test specimen against different surfaces.

A major advantage of drop testing is that the testing can be developed to mimic actual end-use conditions, and materials can be evaluated in that context. One disadvantage is that drop tests often require fully assembled products. While this can provide valuable feedback to confirm the validity of a given design, drop testing rarely provides any useful information in the early phases of design and material selection.

5.6.9 Tumble Testing

Tumble testing is another type of impact testing, which involves a number of different parameters. In a typical tumble test, one takes a number of test specimens and places them in a closed container. Quite often, other items will also be placed in the container, such as rocks, small pebbles, and abrasive media. The container is then placed in a test machine which will tumble the container, in a more or less random manner, for a period of time. In the process, whatever is inside the container will be subjected to a wide variety of low-speed impacts, in a diverse number of loading conditions (in common English, the test specimens are in for a jolly good ride).

At the end of the test, the container is opened, and the test specimens are inspected and evaluated. While the evaluation may involve quantitative measurements (such as weight loss due to pieces being broken off or abraded away, or the change in measured value of a specific performance parameter), qualitative assessments are often more valuable. Qualitative assessments may include a description of common failure modes, observations of high wear areas, and commentary on overall performance.

A major advantage of tumble testing is that evaluation can often be done very quickly. Unlike other impact test methods, where test specimens are evaluated in a solitary manner and the results are then summarized, tumble testing is typically done on a lot basis, where a number of given samples are tested together. As such, tumble testing can often provide the fastest feedback in a trial-and-error type of evaluation. In this type of scenario, a number of samples are prepared and tested en-masse, using a given set of test parameters (which normally mimic actual end-use conditions). The entire lot is then evaluated in a collective manner, and quickly evaluated. Based on the results, the materials can be changed, and/or the test parameters can be adjusted, and another lot can be prepared and tested.

A disadvantage of tumble testing is that it can be difficult—and sometimes impossible—to obtain quantitative data on specific individual parts. However, tumble testing is rarely used for this type of analysis.

The beauty of tumble testing is that it is an all-encompassing test. It is a rock-and-roll, take-no-prisoners, sink-or-swim, do-or-die kind of test. It is the engineering equivalent of playing No-Limit Hold 'Em poker, looking at your cards, evaluating the bets on the table, and then declaring, *I'm all in* (Tables 5.1–5.2).

5.7 But Is It Tough Enough?

In a practical matter, the term toughness is often used to describe the ability of a material to withstand abuse. After all the engineering analysis is said and done, very rarely are we concerned with quantifying exactly how tough a material is under 27 different test methods. What we really want to know is whether a given material will provide the desired performance in the end product for a prescribed amount of time. Sometimes, this involves making some tough decisions, based on a combination of published property data, and a mix of custom test methods (some very precise, some not). The two key questions in this process: *What kind of toughness do I need?* and *Is this material tough enough?* (Figure 5.15)

Table 5.1 Comparing Various Impact Tests

Test Type	Test Details				Rate of Impact (Velocity)					Useful for
	Shapes	Complexity	Machine Cost	Cycle Time	mm/s	in/s	km/h	miles/h		
Standard Tensile	Test bar	Medium–high	$$–$$$	Minutes	0.08/0.8	0.003/0.03	0.00029/0.0029	0.00018/0.0018		Basic property data
Izod (ASTM D256)	Test bar	Low	$–$$	Seconds	3500	138	12.6	7.83		Material comparisons, evaluating notch sensitivity
Un-notched Izod	Test bar	Low	$–$$	Seconds	3500	138	12.6	7.83		Material comparisons, evaluating notch sensitivity
Charpy (ISO 179)	Test bar	Low	$–$$	Seconds	3800	150	13.7	8.5		Material comparisons, evaluating notch sensitivity
Gardner[a]	Discrete parts	Medium	$–$$	Seconds	5970	235	21.5	13.4		Material comparisons, impact fatigue
Falling weight[a]	Discrete parts	Medium	$–$$	Seconds	5970	235	21.5	13.4		Material comparisons, impact fatigue
Instrumented	Any	Medium–high	$$–$$$	Minutes	Varies	Varies	Varies	Varies		Comprehensive analysis

Continued

Table 5.1 Comparing Various Impact Tests—cont'd

Test Type	Test Details					Rate of Impact (Velocity)				Useful for
	Shapes	Complexity	Machine Cost	Cycle Time		mm/s	in/s	km/h	miles/h	
High-speed tensile	Test bar	Medium–high	$$–$$$	Minutes		10,000	393.7	36	22.4	Tensile impact at high rates of loading
Projectile	Any	High	$$–$$$	Minutes		Varies	Varies	Varies	Varies	Comprehensive analysis, real-life simulation
Drop[a]	Complete assemblies	Medium–high	$–$$	Minutes		5970	235	21.5	13.4	Qualitative analysis, real-life simulation
Tumble	Complete assemblies	Medium–high	$$	Minutes		Varies	Varies	Varies	Varies	Comprehensive analysis, real-life simulation

[a] Velocity at impact after a 6-ft drop.

5: MATERIAL SELECTION BASED ON PERFORMANCE 179

Table 5.2 Comparing Velocities

Object/Animal	Description	Velocity				Compares to
		mm/s	in/s	km/h	miles/h	
Garden snail	Average speed	4	0.2	0.02	0.01	50× faster than tensile test
Galapagos tortoise	Walking, typical speed	90	3.5	0.32	0.20	
Mouse	Common house mouse, running	3584	141.1	12.90	8.00	Izod test
Roadrunner bird	*Geococcyx californianus*, top speed	8960	352.8	32.26	20.00	Faster than 6-foot drop
African elephant	Charging bull elephant, enraged	11,200	441.0	40.32	25.00	High-speed tensile test
Usain Bolt	World's fastest human, top speed	12,455	490.3	44.84	27.80	
Sparrow	Eunladen European, in flight	14,023	552.1	50.48	31.30	
Sparrow	English, in flight	17,473	687.9	62.90	39.00	
Cheetah	World's fastest land animal, top speed	31,361	1234.7	112.90	70.00	
Toyota Prius	Downhill, following wind, on a good day	45,250	1781.5	162.90	101.00	
Baseball pitch	Fastest recorded velocity	47,042	1852.0	169.35	105.00	

Figure 5.15 So you think you are tough?

5.7.1 Are You Ready to Rumble?

One of my first experiences with evaluating the toughness of thermoplastic materials came when I was working as a product development engineer for a small company named Kransco Manufacturing. Kransco made a number of recreational products, mostly in the swimming and surfing area. They made floating swimming pool lounges, water basketball games, even boogie boards. One of their products was a knee board sold under the brand name Hydroslide. The user would kneel on the board, and then get towed behind a boat. It was similar to water skiing, but much easier, and a whole lot more fun.

The Hydroslide was a rotationally molded product. I had been to a seminar on rotational molding, and had learned a few things about the process and the materials that were used, and we decided to explore some different materials in this product. Among other things, we wanted to evaluate how each of these materials performed during the rotational molding process. Were they easier to mold? Were the parts stiffer or more flexible? Was the color more vibrant? Did they assemble easier? So we had some samples made in various versions of polyethylene—LLDPE, HDPE, etc. We then went about evaluating them.

One of the things we discovered was that there were some subtle differences in the surface characteristics of each version. Sometimes this was helpful, as it made easier to attach the foam knee pad to the board, and sometimes it was not. We also began to notice differences in their durability. In an attempt to quantify this, my boss and I hooked up a bunch of samples behind his car, and we drove around the parking lot for 15 min or so, towing the boards behind us, and doing everything we could to make them tumble and spin and bounce and collide. While there was no alcohol involved—and no animals were harmed in the testing—we did have a lot of fun. And we were also able to determine which materials had the best durability.

A day or so later, I heard someone in the company—probably one of the bean counters—had seen us driving around the parking lot and had marched into the president's office, and said something to the effect of, *Do you know what those idiots in product development are doing right now? They are driving around the parking lot, towing a bunch of plastic junk behind them.* He looked up from his desk, looked her straight in the eye, and responded, "They are doing exactly what they need to be doing." We never had a single complaint about our test methods after that.

5.7.2 Cutting the Grass

In the mid-1980s I found myself working for a small company named Allegretti and Company. They made a line of motorized lawn and garden tools that were sold by Sears. Their main products were leaf blowers, edge trimmers, and weed wackers (Figure 5.16).

The performance requirements of these products were similar to traditional power tools, with one additional requirement. They had to withstand projectile impacts—at a high velocity. These tools all had high-speed motors with whirring blades and knives and fans. Any of those items could make impact with a stone, or a piece of steel, or a chunk of wood, and

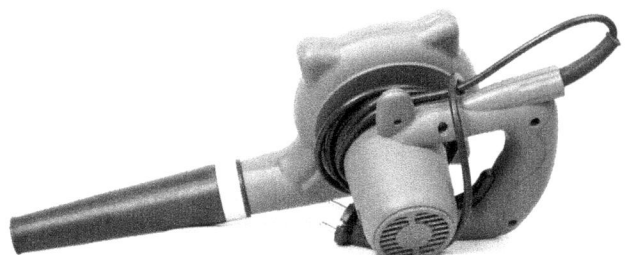

Figure 5.16 Leaf blower. momopixs/Shutterstock.com.

then propel it outward at high velocity. Hopefully, it would either hit the ground, or hit the protective shroud and then exit the exhaust chute.

The shrouds and exhaust chutes were made from tough materials like polycarbonate, ABS, occasionally even polypropylene. The key was projectile testing, specifically small objects weighing grams or ounces that were projected at fairly high speed. The performance requirements might vary, minor deformation allowed, no structural failure allowed, etc. While projectile testing can be a slow and tedious process, it can provide valuable information on performance in real-world applications.

Sometime in the early 1990s, a DuPont colleague of mine went to a trade show for the lawn and garden industry. This was a manufacturer's show, where new equipment was showcased. It involved not just weed-wackers, but lawnmowers and chain saws and snowblowers, not just for consumer use, but commercial equipment as well, even heavy machinery used in agriculture. GE Plastics had an exhibit at the show, promoting the use of their materials. They were always very good at marketing.

They had gotten an injection mold made, a very large injection mold, and had molded a lawnmower deck. Not just the deck of small four-wheel push-type lawnmower, but a big industrial lawnmower. The blades under these decks can turn rocks and other objects into deadly weapons, so the decks themselves needed to be able to withstand a tremendous amount of impact energy. They had one of these decks in a protective enclosure in their booth, hanging sideways like a giant gong. Every hour, on the hour, they fired a projectile from a cannon, and it would strike the deck with tremendous velocity. The deck would ring like a giant gong, reverberating throughout the convention hall, while the projectile would simply fall to the floor.

While some might think that the purpose of this demonstration was to prove the toughness of thermoplastics from GE Plastics, I prefer to think of it a subtle reminder of importance of projectile testing when evaluating the use of thermoplastics in an application subject to high velocity impacts.

5.7.3 Chicago Style

In the mid-1990s I relocated to Chicago. Chicago is a great American city. Not only is it an alpha global city [4], it is in many ways the epicenter of manufacturing in the Midwestern United States. Or as I like to describe it, Chicago is the buckle of the rust belt.

There are hundreds of manufacturing companies in the midwest, making everything from auto parts to zebra printers. Many of these companies use thermoplastics in their products, sometimes to reduce costs, sometimes to reduce weight, but often times because they can achieve a level of performance due to the toughness of a specific thermoplastic. Many of these companies have their own test labs, where they develop testing protocols that are tailored to their unique application.

Imagine you are working for a company like Whirlpool and are tasked with selecting a material to be used in the drive train of washing machine. And you know that many of the components in that machine are covered under an extended warranty, lasting 3, 5, perhaps even 10 years. Are you going to select a material based solely on published values of Izod test data? I doubt it. You are going to develop some prototypes, choose some candidate materials, and put them in your test lab and test them for days, weeks, months, perhaps even years. If, or more likely when they fail, you are going to look at them under a microscope and evaluate the failure mode. Or you may take high-speed video to see what is happening dynamically during the moment of impact.

Or imagine you are working for a manufacturer of power tools, making drills, saws, sanders, grinders, nail guns. Are you going to select a material based on Gardner impact test results? Or are you going to do more in-depth evaluation?

In many of these kinds of situations, the selection of a material with the right amount of toughness is dependent on the performance level of the end product. The level of performance that is required is very often related to the sales price of the end product, which in turn affects the expectations of the consumer. No one expects a cordless drill that sells for $19.99 to have the same level of performance as one that sells for $199. However, if you went and bought a $199 drill, and then the first time you used it, you dropped it onto a concrete floor, and it broke, you probably would not be very happy.

So companies that make these kinds of products devise all kinds of performance tests, dropping them from the roof onto concrete, running them over with a truck, dragging them through mud, doing everything they can to abuse them, and evaluate their toughness.

Now one might say, "These tests are statistically invalid, the measurements are inaccurate, and the results are not reliable and repeatable. Even if they are performed under controlled conditions, they have no correlation to the actual in-use performance of the material in the intended application."

To all of that, I will readily agree. However, I will also argue that these kinds of tests represent real-life, in-use scenarios, and if the product—and the materials that are selected—cannot pass these tests, then you have no business using those materials in the application.

The 1987 movie *The Untouchables* featured Kevin Costner playing Elliot Ness, the federal agent tasked with bringing down the famous Chicago gangster Al Capone. In one scene, a Canadian Mountie chastized his interrogation techniques. "Mr. Ness, I don't approve of your methods." His response: *Oh yeah? Well, you're not from Chicago.*

5.7.4 The Thrill of Victory (and the Agony of Defeat)

Throughout history, people have always played sports. And just as with other man-made things, the materials that have been used to make sports equipment have changed throughout history. Today, while sports equipment is still made from traditional materials like wood, leather, and metal, thermoplastics are increasingly used in a diverse numbers of ways. In some cases, the sporting goods industry is at the forefront of plastics engineering.

Just as in other industries, thermoplastics are used to reduce weight and decrease the cost of manufacture. They are also used specifically to improve product performance. And while strength and stiffness and chemical resistance have always been an important part of performance, the toughness of the material is often a critical parameter.

One important material is polycarbonate, which is used in all kinds of applications, ranging from football helmets to face guards and safety glasses. The toughness of polycarbonate is outstanding and it excels in all aspects of impact testing, especially with Gardner-type impact and projectile testing (Figure 5.17).

Another important material is nylon, which is used in rope and webbing, and also in high-performance sports fabrics. In these the toughness requirement is in pure tension. Tensile impact testing and high-speed tensile testing are useful in these applications. The elongation of the material is often equally important.

Thermoplastics are often used in high-performance zippers on backpacks and tents and sports clothing. These not only require tensile toughness, but often low-temperature toughness as well. Finding impact data at very low temperatures can be challenging. Thorough product testing is often required.

Equipment in many sports often has specific structural requirements, combined with a need to withstand impact and abuse. Whether this is the frame that supports the basket on a lacrosse stick, or structural support for the wheels used on an inline skate, these need to have the big three: strength, stiffness, and toughness.

5: Material Selection Based on Performance 185

Figure 5.17 Football helmet. pbombaert/Shutterstock.com.

Ironically, the toughness requirement for the frame of an inline skate is similar to the carrier on a nail gun. They are both subject to repeated impact over long periods of time, with occasional high-energy impacts due to drops or collisions. It is no coincidence that toughened, glass-reinforced nylon is used in both of these applications.

One interesting aspect of sporting equipment is that most products are rarely used in a continuous manner for extended periods of time. This is very different than most industrial applications, or even power tools. While a sports product may be expected to last for several years, it is typically only used for a short period of time, perhaps a few minutes, perhaps a few hours or days. The product may then be stored in a garage, or outside in the sun, even partially submersed in water. However, the next time the product is used, the same level of performance is typically expected. As a result, evaluating the toughness of a material for a given application often means accounting for exposure to temperature, chemicals, radiation, and time (remember the Four Horsemen?). Quite often, property data is simply not available, and extended life testing of the actual product is required.

5.7.5 Dealing with JRA

Another classic example of toughness in the sporting goods world comes from the bicycle industry. SRAM was a manufacturer of bicycle components, which they primarily sold to bicycle manufacturers, but they also sold parts in the aftermarket. Their primary product was a rotary shifting mechanism known by the brand name GripShift. It consisted of a rotating plastic cylinder that acted as a cam to pull and release the shifting cable in a precise manner. It was a very simple solution to a complex problem.

SRAM had recently started manufacturing a rear derailleur. A derailleur is a mechanism that moves the chain from one set of sprocket teeth to another, allowing a bicyclist to change gears. The rear derailleur is a complex mechanism that involves a parallelogram, several springs, and a couple of pulleys. It not only allows for changing gears, but also provides for proper chain tensioning through the entire range of gearing. SRAM had developed a design that allowed for more precise positioning of the chain to the individual gear teeth. An SRAM derailleur was used to win the gold medal in the first Olympic mountain bike race in Atlanta in 1996.

There were two main parts in the derailleur that were made of plastic. Known as the knuckles, they were injection molded of a reinforced grade of nylon. The reinforcing agent was a type of ceramic, and the resulting parts were incredibly stiff, allowing the mechanism to be extremely precise. Unfortunately, SRAM begin experiencing a high rate of returns due to field failures, known as FFR. I was asked to investigate the issue and determine what could be done.

Rear derailleurs take a tremendous amount of abuse, especially in mountain biking. Riders hit rocks and tree branches and stumps, and occasionally, these bumps will force the derailleur directing in the spokes of the spinning rear wheel, at which point all hell breaks loose. We were getting returns of broken derailleurs showing all kinds of different failure modes. However, one common description of the activity was described as *Just Riding Along*, or JRA. It is the bicyclist's version of the classic excuse, *I wasn't doing anything wrong*. Replicating a field failure from JRA was practically impossible. However, it was easy to replicate failure modes from a side or frontal impact.

The theory we developed was that these derailleurs were failing due to impact fatigue. The rider would be out for a ride, hit something hard, but keep on going. Sometime later, maybe the next time out, they would put a little force on the chain and the derailleur would snap. That extra force was the proverbial straw that broke the camel's back.

We began exploring the use of different versions of nylon, with different reinforcing agents and different toughening additives. Our test methods involved not just pendulum-type impact tests, but also drop tests, with different weights dropped at different heights. Our goal was not just the highest possible impact strength, but the ability to withstand a high number of impacts at lower impact energies. We eventually settled on an aromatic polyamide (a type of nylon), with 50% glass reinforcement, and a

5: Material Selection Based on Performance 187

Figure 5.18 A severely damaged rear derailleur.

proprietary toughening agent. This material was TOUGH. Ironically, it was a material that was being widely used in other applications in the sporting goods industry, including bindings on snow skis. It was stiff, it was strong, and it was unbelievably tough.

The FFR dropped substantially, and several months later one of the field technical reps dropped a broken derailleur on my desk. It was a derailleur that had gotten into somebody's rear wheel. The aluminum connecting links were severed, the aluminum plates that held the pulleys were twisted and bent, but the plastic knuckles were more or less intact. Now that's toughness (Figure 5.18).

5.7.6 The Shupe Test

SRAM also had plans to manufacture a set of brake levers made 100% from plastic, using the same material that worked so well in the derailleur knuckles. Jeff Shupe, the Vice President of Manufacturing, was not in favor of making a plastic brake lever. He used to stroll into the R&D lab every now and again, grab a prototype brake lever, and then twist, pull, yank it every which way until it broke. He would then toss the broken brake lever on the work bench and walk out, without saying a word. The implication was clear. If he could break it, what was going to happen in the real world?

So the engineering team went all out in their efforts to make a robust design, knowing we had a material that was stiff enough, strong enough, and tough enough. Along with other performance tests, we also created a test where we would mount the brake lever assembly to a solid steel bar, and then bolt the steel bar to a mounting plate. We would then have Jeff come in, and he would first wrap a towel around his hand (kind of like how a prize fighter would tape his hands before a big fight), put on a set of safety goggles, and then grab the lever and do everything he could to try and break it. We called it "The Shupe Test." We eventually were able to create a design that he could not break. The design went into production a few months later.

Each year there is a huge trade show for the bicycle industry, called Interbike. We decided to bring "The Shupe Test" to the show. For those of you who are serious cyclists, you can appreciate the level of testosterone that exists at a show like this. When prospective buyers came to the booth, we would casually show them the new plastic brake lever, and when they expressed doubt that a plastic brake lever could work—let alone stand up to the abuse that a mountain bike endures—we would invite them to take the Shupe test. *Go ahead, grab it. Bend it. Twist it. Break it. Come on, can't you break it, what's the matter, are you a pussy?* Inevitably, they would work themselves into a frenzy trying to break it, only to eventually give up, and say something to the effect of, "I can't believe how tough that is." SRAM sold a lot of plastic brake levers that year.

5.7.7 The Bottom Line on Toughness

The toughness of a given material cannot always be evaluated by simply looking at the published values of one specific test. When it comes to material selection, toughness should always be evaluated using a variety of different methods. The methods should include a mix of quantitative and qualitative criteria, including tests that can be correlated to the actual in-use performance of the final product.

I remember a quote from Bill Miller, a mutual fund manager who ran a fund named Legg Mason Value Trust (LMVTX), which at one point had the distinction of beating the annual return of the Standard & Poor 500 Index for 15 consecutive years. (Full disclosure: I invested some money with him for some of that period, thank you very much.) The quote said something to the effect—"At LMTVX, we use a multi-factorial value analysis approach. That is, we seek to evaluate real value using a variety of different methods [5]." I would encourage a similar multifactorial analysis approach for thermoplastic material selection, especially when it comes to evaluating toughness (Figure 5.19).

5.8 Surface Properties

Another key set of mechanical properties has to do surface characteristics of the material, specifically with how the material interacts with another material when they are in motion relative to one another. These

5: Material Selection Based on Performance

Figure 5.19 Selecting materials based on toughness. totallyPic.com/Shutterstock.com.

characteristics include friction, lubricity, and wear, the study of which is known as tribology. Other surface characteristics include hardness.

5.8.1 Friction

Friction is the force that provides resistance as two objects slide against one another. It can be measured in many different ways, the two primary ones being static (stationary) and dynamic (moving) friction. It is typically expressed through the coefficient of friction, which is a dimensionless ratio comparing the sliding force to the normal force. While friction is an inherent material property, it varies depending on what material it is measured against. Material pairs with a high coefficient of friction are considered sticky, while those with a low coefficient of friction are considered slippery. Thermoplastic materials are selected for either reason, depending on the application. For teflon-against-teflon (PTFE), the coefficient of friction is 0.04, one of the slipperiest combinations known to man.

5.8.2 Lubricity

The lubricity of a material is not a specific property. Rather it is an assessment of a material in a given system. Many thermoplastics have inherently excellent lubricity in a wide range of environments, and there also additives which can be used to enhance the lubricity. Friction and lubricity are often related, and while data for coefficients of friction is frequently published, data is less available for lubricity.

5.8.3 Wear

Wear describes the change in a material as it in motion against another material. It includes permanent deformation due to yielding, loss of material due to abrasion, surface changes to crack propagation, and erosion, which is a loss of material due to the cutting action of liquids or small particles.

5.8.4 Hardness

Hardness is the resistance of a solid material to permanent deformation when localized forces are applied. It is a form of compressive strength. It is commonly measured via scratch hardness and indentation hardness. Scratch hardness, as the name implies, is the resistance to a sharp object as it moves against the material's surface. Soft materials will readily deform, and a gouge or scratch will be left behind. The size and depth of the scratch, the appearance of the scratch (smooth or rough), are all elements of scratch resistance.

Indentation hardness involves taking an object, often a hardened steel ball, and pressing it against the material with a defined force, and then measuring the depth of the resulting indentation after the ball is removed.

In general, most thermoplastics are softer than your typical metal materials. Elastomers and foams also have very different behavior when it comes to hardness, since they inherently have a high amount of elasticity, and can be easily compressed without having any permanent deformation. Foam hardness is often measured based on the amount of compression during a given load. Known as indentation load deflection, this measures the ability of a foam to support a given weight.

Selecting thermoplastic materials based on their surface properties typically needs to be done in context of the system that the materials are going to be used in.

As an example, thermoplastics are frequently used in rotating applications, either as pulleys or bearings or cams. The performance of a specific thermoplastic for a given application will be heavily dependent on the speed of rotation, and the forces involved. It will also depend on the material used for the shaft, how hard that material is, and what surface finish it has. Rather than trying to find the perfect thermoplastic for that bearing, it may be simpler to focus efforts on the shaft itself.

5.9 Key Electrical Properties

When we think of the electrical properties of a thermoplastic, we often think of their ability to work as electrical insulators. That is, they are *not* expected to conduct an electric current, rather, they are expected to *prevent* any current flow.

5.9.1 Insulating Plastics

Most thermoplastic materials are inherently nonconductive and work well as electrical insulators. There are numerous test methods for evaluating the insulating properties thermoplastics, including dielectric strength, arc resistance, surface resistivity, volume resistivity, and dissipation factor.

Dielectric strength is the ability of an insulating material to withstand electric stress without breaking down. It is dependent on not just the material, but the thickness, the rate at which voltage is applied, along with a number of other factors. Just like other measurements of strength, the dielectric strength of a given thermoplastic is affected by temperature, exposure to radiation and chemicals (including water and humidity), and time.

Selecting thermoplastic materials for insulation purposes is a complex endeavor, and well beyond the scope of this book. A list of suggested references is provided at the end of this chapter for readers seeking more detailed information.

5.9.2 Conductive Plastics

At the opposite end of the spectrum is the issue of electrical conductivity. While most thermoplastics are inherently nonconductive, there are a number of additives that can be used to modify the base resin and enhance the electrical conductivity. While these modified resins will never achieve the level of electrical conductivity of most metals, they can provide *some* conductivity, making them useful for applications requiring antistatic performance (to prevent dust buildup and/or static cling), control over static discharge, and sometimes even for electronic shielding.

The selection of the thermoplastic material itself is usually driven by structural requirements first, and then the type and amount of conductivity required will dictate what additives are to be used. The use of conductive plastics is a specialty field, but is growing rapidly. Often times, a specialty compounder can provide in-depth technical guidance.

5.10 Properties of Form

One of the unique characteristics of thermoplastic materials is that they can used in a wide variety of manufacturing processes. Each of these processes is unique, with advantages and disadvantages, along with a unique set of constraints.

In a similar manner, every thermoplastic material is also unique. Not just in terms of its measured physical properties, but also in how it responds to each manufacturing process. The ability of a material to be shaped and formed into a final product that satisfies (and hopefully delights) the needs of the end user is sometimes even more important than the properties of the material itself. Unfortunately, this aspect of thermoplastic material selection is often overlooked.

5.10.1 Size

Products come in all shapes and sizes, as do plastic parts. Selecting a thermoplastic material based on the part size may involve evaluating several properties, as well assessing whether a material can be used in a specific manufacturing process.

At one end of the scale is the question, *How big can we make the part?* This is not only a question of part volume, but for many processes, there is a related question, *How thick can it be?*

At first glance, one would think that the maximum size of a plastic part is only limited by the size of the processing equipment being used. If you want bigger parts, you just get bigger machines. However, every process has practical limits. One of the responsibilities of a good part supplier is to understand these limits. While performance is always important, quite often the selection a thermoplastic material for use in very large parts is based on the experience a supplier has in working with that material.

One general guideline is to select materials that have good thermal stability at the processing temperatures involved. Since you have large parts, you will have a large volume of material, and the cycle times will be longer. So, the material will probably be at elevated temperature for a longer period of time than if you were making smaller parts (Figure 5.20).

In terms of how thick one can make a plastic part, it is highly dependent on the manufacturing processes. Parts made via the structural foam process can be significantly thicker than parts made via injection molding or rotational molding. But even with injection molding, it is possible to mold parts that are ½″–¾″ thick (12–18 mm).

5: MATERIAL SELECTION BASED ON PERFORMANCE 193

Figure 5.20 Herman Miller Equa chair shell, injection molded in one piece from 30% glass-reinforced PET.

As a general rule, it is often easier to injection mold very thick parts using amorphous resins. This is not because of fill, but because of differential shrinkage. With semicrystalline materials in thick cross-sections, the rate of crystallization and the total amount of crystallization can vary significantly from skin to core. As a result, the total amount of mold shrinkage can vary through the part. This can lead to distortions (warping and/or sink marks).

At the opposite end of the scale the question becomes, *How small of a part can we make?* Also, *How thin can it be?* It is possible to injection mold very small parts, sometimes as small as the head of a pin. Known as micromolding, material selection for this technology is highly specialized. References are provided at the end of this chapter for readers interested in this technology.

In terms of how thin one can make a plastic part, the question usually involves ratios, as in how thick the part is when compared to how long and wide the part is. In injection molding applications, we are often concerned with the minimum wall thickness of part. This wall thickness is a limiting factor in the fill of the cavity during injection. In other processes, the limiting factor in lower thicknesses is structural integrity, both during processing and in the final part.

In the Equa chair shell shown earlier, the ratio of part thickness to part length is not unusual. Part fill typically only becomes a problem when

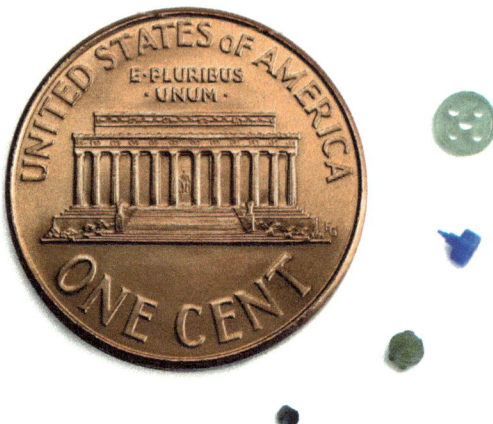

Figure 5.21 Micromolding.

the ratio gets very small—either because the part is very thin, or the part is very long. One useful property in evaluating this is spiral flow data, which involves injecting the resin into a cavity in the shape of a spiral. The length of the amount of material that will flow before cooling, along with the overall volume of material injected, is an excellent indicator of how well the material flows. This property is much more useful than Melt Flow Index (MFI), which is a popular (and often inaccurate) way of measuring resin viscosity.

In general, materials with low viscosity have a high MFI. Often, resins are specifically formulated for high flow. These version are frequently designated HF (for high flow). However, in molding very thin parts, there is a practical limit, even with a HF resin. A good part supplier will understand these limits (Figure 5.21).

One general guideline is that semicrystalline materials can typically be used in thinner wall thicknesses than amorphous materials. Since they have a distinct melting point, they easily transform into a low viscosity liquid for the injection phase, and then rapidly solidify in the mold.

5.10.2 Shape

The performance of many products is based on their overall shape, not just in terms of appearance, but in terms of function as well. The shape and form of the plastic parts in that product play an important role. While some shapes are a function of the processing technology used, such as a blow molded beverage container, sometimes shapes are based on the unique characteristics of a given material. However, this information is rarely captured in material data sheets.

As an example, with many processes it is a standard practice to maintain a uniform wall thickness. There are many reasons for this, including a desire to have uniform shrinkage over the entire part. However, there are some processes, such as structural foam and gas-assist injection molding, that allow for high variations in the wall thickness of the final part. This capability of molding both thick and thin sections in the same part can provide performance advantages in regards to shape. There are also some thermoplastic materials that defy convention and can be injection molded with both thick and thin sections in the same part, for example, some of the polyester block copolymer elastomers (known by the trade names Hytrel® and Lomod®).

Another aspect of shape involves tolerances, specifically dimensional tolerances on the final shape. Fit and function are almost always dependent on dimensional tolerances, and controlling them in the final part can be an important criteria for performance. Quite often the part supplier is responsible for dimensional tolerances, however, part design and material selection play a critical role.

Dimensional tolerances for plastic parts include not only flatness and straightness, but dimensions on features of size and positional tolerances as well. In order to achieve high precision, it is common practice to utilize low-shrinkage materials (amorphous resins). The line of thinking is that with less shrinkage, there will be less variation. There is some validity to this. There are others that argue dimensional variation in the parts is due to variation in the manufacturing process itself, and with proper process control the dimensional variations can be minimized. There is some validity to this argument as well.

As a general guideline, it always makes sense to select a material that is process friendly, that is, one that allows for less than perfect processing conditions. This may mean using an amorphous material instead of semicrystalline material, or a material that is a blend of both. It may mean using a mineral-filled material instead of glass fiber-reinforced material (in general, mineral-filled materials shrink more evenly, which can help reduce warpage and other dimensional variations).

5.10.3 Appearance

Quite often, the appearance of a plastic part is a critical performance factor. While this often involves color, it also involves texture and gloss.

When we think of the texture of a plastic part, we often think of the texture that was on the surface of the mold, which has been transferred to the surface of the part. We often forget that the textured surfaces in the part are NOT a perfect copy of the texture that was in the mold. There will be surface imperfections, minor deviations here and there, etc. Even if the mold was polished to the highest standards for smoothness, the molded plastic part will NOT have that level of smoothness. Furthermore, I would postulate that every thermoplastic material will have its own unique signature in how it replicates the mold surface. I think of this as the material texture. Unlike metal fabrication, which often involves secondary finishing processes, thermoplastic materials are rarely buffed and polished and honed. So the surface texture on the final product is typically an "as fabricated" texture, and it is dependent on the material as much as the manufacturing process. As far as I know there is no published data on this, and very few people are even concerned with it.

On the other hand, in many applications, the gloss level of the material is an important parameter and materials are often screened for gloss. Some materials, like ABS and polycarbonate, just seem to have an intrinsically high gloss level, and they come out of the mold with very high gloss, looking shiny and bright. Other materials, like polypropylene, seem to have a lower gloss level, more like eggshell. They come out of the mold looking a bit flat. However, unlike the paint industry, which has measurement data on gloss levels every which way can you can think of, resin suppliers rarely provide information on the gloss level of thermoplastics. This kind of information is loosely held, mostly as a matter of tribal knowledge.

Then there is the topic of color. Color is a complex subject, and well beyond the scope of this book. However, at a basic level, we need to understand that the appearance of an object involves not only the color of the object (the wavelengths of light that are being reflected), but the gloss level and texture of the surface that is reflecting the light. Anyone who has ever held a textured plastic color chip can tell you that the appearance changes as the texture changes over the object's surface.

Every thermoplastic material has its own innate color. Some are clear, some are a milky white, others are brown, some are even black. (A thermoplastic material in its innate color is referred to as natural, regardless of the exact color.) Furthermore, every thermoplastic material responds to pigments in a different way. These pigments—whether they be dyes, powders, flakes—are what determine the final color of the fabricated part. However, how the material responds to the pigment is a critical aspect of the overall appearance. Is the color rich and vibrant? Or is it dull and muted? Is it smooth and glossy? Or rough and eggshell?

Selecting a thermoplastic material for optimal appearance should include considerations for not only color, but gloss and texture as well.

Figure 5.22 Material samples showing some standard colors.

Unfortunately, this type of information is rarely contained in property data sheets or resin brochures. Instead, it is subculture all its own, with a language and jargon that can make your head spin. Those interested in this area are encouraged to investigate the resources listed at the end of this chapter (Figure 5.22).

5.11 Some Final Guidelines

I have been working with thermoplastic materials for over 30 years now. Throughout my career I have been complaining that most people go about selecting thermoplastic materials for performance the wrong way (and I tried to cover some of those wrong ways in this chapter).

When I started writing this chapter, I thought it would provide me an opportunity to set the record straight, and tell the whole world, *Here is how it should be done.* I was going to create a flow chart, a decision matrix, and add some interesting stories on combining technical data, real-life experience, and touchy-feely stuff into a magic elixir. This elixir would provide a fool-proof methodology for material selection, a method that would work every time, for any application, no matter what. Sadly, I have not found the magic elixir.

The reality is that material selection is a complex process [6]. It is not like ordering dinner. *And for you, sir?* "I'll have the generic prime nylon, type 6 please, natural color, with just a touch of glass reinforcement." *And how much toughness would you like with your nylon?* "Standard toughness is fine, thank you." (Additives always give me a headache.)

The material selection process is kind of like playing a pin ball machine. You start flipping through the process, adding up the points, when all of a sudden the warning bells go off, lights start flashing and jinga-jinga-jinga-jinga, the next thing you know you've dropped the ball, and have to start all over.

Fortunately, using the information you have gained in this chapter, you can work through this process with knowledge and confidence. There are a few tools that I would recommend you utilize.

5.11.1 Conceptual Tools

On a conceptual level, approach this process not as a singular activity, but as a collective effort. While the design team often has the responsibility for material specification, the sourcing team plays a major role as well, which is not just in negotiating supply contracts with material suppliers and/or distributors, but in managing the overall logistics of the supply chain as well. One could specify the most perfect thermoplastic material in the world, but it will not improve the performance of your product one iota if that material cannot be sourced.

The manufacturing team also has an important role. This includes the tool makers, the molder, even the people on the manufacturing floor. They are the ones who are going to be working with the material, helping to convert it from a bag of plastic resin into a functional, viable product. They will often have first-hand experience with the materials and will have knowledge and insights on all kinds of things that cannot be found on property data sheets.

In your process of material selection, do everything you can to utilize the collective knowledge and experience of your team. This can be a formal or informal process, but it should involve face-to-face conversation with the members of your team. We live in a digital age, where data and information always seems to be at our fingertips, if only we know the right terms to use in a search engine. And while databases and online resources are great, nothing beats talking through the issues with your team.

Finally, use your imagination. As we have discussed throughout this book, thermoplastics are unique materials, with unique properties, and unique behaviors. Humans have been working with metals for hundreds of years, yet we have been working with plastics for less than a century. If we are to use them effectively, we need to use our imagination.

> I am enough of the artist to draw freely upon my imagination. Imagination is more important than knowledge. Knowledge is limited. Imagination encircles the world.
>
> **—Albert Einstein [7].**

5.11.2 Mathematical Tools

There are a number of mathematical tools that are used in the world of manufacturing. Statistical Process Control is used to monitor and control the process parameters that are used to manufacture parts. Tolerance Analyses are used to analyze and predict fit and function of final assemblies. Pareto Analysis techniques are used to assess the contribution of various factors in problem situations. Design of Experiments techniques are used to quantify the variables in a given process or application.

Most of these techniques are quite rigorous, and the mathematics are precise and exact. Most of them require measurement data that is based on manufactured parts. In other words, while you can discuss them in concept, to utilize the techniques you have to have actual parts.

In the world of material selection, ideally we want to select a material *before* any parts are made. In this situation we will not have any measurement data, and the mathematical tools discussed above are not applicable. While there is published data on measured properties of materials—lots of published data—there are very few mathematical tools available to evaluate this data in a comprehensive manner. One can compare the tensile strength data of different materials, or the flexural modulus, or the heat deflection temperature (HDT); it is difficult to compare all of the available data on material candidates and be certain you are making an optimal selection. However, there are mathematical tools that can help guide this process.

In the field of failure analysis, there is a methodology known as Failure Mode and Effects Analysis [8] (FMEA). FMEA involves a systematic approach to failure analysis, where individual components, subsystems, and complete systems are evaluated for possible causes of failure. The effects of each failure are then assessed and assigned a numerical value for severity. The possible causes are assigned a probability of occurrence (what percent of the time they might occur), as well as a detection rate (how often they will be detected before failure). The severity value is multiplied by the probability to give a rating of criticality. Items with a high level of severity and a high chance of probability are the most critical.

The criticality is then multiplied by the detection value to give a Risk Priority Number (RPN). The numerical value for detection is inverse to the detection rate (a detection rate of 100% has a numerical value of zero, while a detection rate of 0% has the highest numerical value). The failure modes are then ranked basked on RPN. Additional information can then be evaluated based on whether root causes are to be evaluated, what mitigating efforts should be made, etc. While FMEA primarily utilizes qualitative assessments, it provides a robust means for quantitative analysis of complex systems.

In a similar vein, I would encourage those engaged in material selection to formally document their selection process using a Material Properties Effects Analysis (MPEA). MPEA involves a systemized approach to material selection, where materials are carefully evaluated for their effect on the system. However, instead of evaluating the effect of failure, the process is used to evaluate the effect of specific material properties. The intent is to determine the critical material properties of the materials that are used in the system.

In an MPEA, we start with a list of possible properties that will be important in the end-use application. This might include tensile strength of the material in component A and the impact toughness of the material in component B. This list can be quite lengthy, depending on the complexity of the system. For each we assign an initial importance to that number, from 1 to 100. (The assumption is that we know exactly how important that property is. We will come back to this.) From there, we itemize factors that could affect that specific material property. These could be processing factors, environmental factors, etc. We then assign a probability of occurrence for that factor. The occurrence is then multiplied by the importance to give a numerical value of criticality.

From here we then begin to make subjective assessments of the factor. For instance, the effects of short-term exposure to heat are often fully reversible. The material temporarily loses strength and stiffness but will regain them once the heat is removed. It does not affect the performance as long as we are not using the part during that exposure. This subjective assessment is assigned a value and is multiplied by the criticality to give an overall assessment. Notes and comments are made to further evaluate the factor.

Often times, we may find that our assumptions about materials and the relative importance of a given property are not based on facts. As an example, we may say tensile strength is absolutely the most important property, and we assign it an importance of 100. We also intend to specify "No Regrind Allowed," because our application is so demanding. However, after performing an MPEA, we realize that a simple solution is to increase the cross sectional area of the structure.

Just like FMEA, MPEA utilizes a number of qualitative assessments, and the overall quality of the process depends on the accuracy and consistency of these assessments. Unlike FMEA, MPEA also allows for quantitative input, based on real-world data. While MPEA is relatively new, when properly applied it provides a robust means for the quantitative analysis of material property requirements in complex systems and can be a valuable tool for material selection (Table 5.3).

Table 5.3 Material Properties Effects Analysis

Property/Importance	Factors			Assessment				Comments	
	Item	Effect	Probability of Occurrence	Criticality	Reversible/ Recoverable	RR Value	Overall Importance	Net Effect	Possible Mitigating Efforts
Tensile strength									
100	Improper drying	Loss of strength	2%	2	10%	0.90	1.8	Brittle failure	Increase cross sectional area; implement process controls
100	Chemical attack	Loss of strength	100%	100	0%	1.00	100.0	Softening, ductile failure	Change material
100	Heat exposure (ST)	Lowered strength	10%	10	100%	0.01	0.1	Temporary loss of strength	Investigate further
100	Heat exposure (LT)	Loss of strength	20%	20	0%	1.00	20.0	Permanent loss of strength	Change material

Continued

Table 5.3 Material Properties Effects Analysis—cont'd

100	Use of re-grind	Lowered strength	100%	100	0%	1.00	100.0	Lowered strength	Increase cross sectional area

Toughness (single impact)

50	Improper drying	Loss of toughness	2%	1	10%	0.90	0.9	Brittle failure	Investigate further
50	Heat exposure (LT)	Loss of toughness	20%	10	0%	1.00	10.0	Brittle failure	Change material

Toughness (repeated impact)

80	Heat exposure (LT)	Loss of toughness	20%	16	0%	1.00	16.0	Brittle failure	Change material
80	Use of re-grind	Lowered toughness	100%	80	0%	1.00	80.0	Long-term failure	Investigate further

5.11.3 Determining Critical Material Properties

One of the major challenges in thermoplastic material selection is in determining which material properties are most important. As stated earlier, in most applications, it is a combination of properties which, when integrated together in a specific design, result in the desired performance. Most plastic parts perform multiple functions as well. It is unusual for a plastic part to perform a single function, such as acting only as a stiffening brace. If this were the case, we would select a material with adequate strength and stiffness and be done with it. However, more often than not a plastic part will not only act as a stiffening brace, but it will also provide a mounting interface to another part, a handle to grab onto, along with a hard stop for a mechanism to engage and disengage. So now we have multiple requirements, each with different evaluation criteria, and each with different demands on the material itself. How do we determine the critical material properties? While an MPEA can provide guidance in this regard, one of the most effective means of answering this question is to evaluate the performance of materials in actual real-life applications.

There is an old adage in the world of financial investment: "Past performance is no guarantee of future results." In the United States, the Securities and Exchanges Commission has even mandated that this statement be included in any marketing document that contains financial performance data. However, in regards to thermoplastic materials, past performance of a real-life product is an excellent indicator of future results. Whenever possible, look at commercial applications of thermoplastic materials and analyze them from a material perspective.

In addition to asking basic performance questions (Does it work? Does it last? Is the product reliable and safe?), investigate the performance limits. If there have been failures, what are the failure modes? Were they due to an unexpected loading condition? A loss of properties due to environmental factors (e.g., chemical attack)? Or due to abuse? These questions need to account for the performance category of the product. For example, a power tool for the industrial/professional market is expected to have significantly higher performance than a power tool for the consumer market. But at the end of this process, you should have a very good indication of the critical properties required in your application.

References

[1] M. Lewis, Moneyball: The Art of Winning an Unfair Game, W.W. Norton and Company, 2003.

[2] E.R. Larson, Big, baad, and beautiful: advances in the use of engineering plastics in the office furniture industry, Reprinted in Modern Woodworking, in: Technical Paper Presented at the SPI Structural Plastics Conference, April 1990, November 1990.

[3] Full disclosure: the author was awarded a Bachelor of Science degree in Aerospace Engineering magna cum laude from the University of Michigan in 1979. He is thoroughly qualified to judge what is—and what is not—rocket science.

[4] "A global city, also called world city or sometimes alpha city or world center, is a city generally considered to be an important node in the global economic system." Wikipedia: The Free Encyclopedia. Wikimedia Foundation, Inc. http://en.wikipedia.org/wiki/Global_city.

[5] K. Kazanjian, Wizards of Wall Street, Wolters Kluwer Law & Business, 2000.

[6] There are many guides for how to choose the best plastic material, compiled by a variety of companies and offering useful tidbits of information. Some of these guides include: ASME. "Metal to Plastic: Design Flexibility." www.asme.org, 2013; Curbell Plastics. Material Selection Guide. www.curbellplastics.com, 2014; Firedrake, Inc. "Choosing the Right Plastics." http://www.firedrakeinc.com, 2014; Norwe, Inc. "Choosing the Right Plastic Material." www.norwe.com, 2008; Quadrant Plastics. "Material Selection Guide."http://www.quadrantplastics.com/na-en/home.html, 2007–2015; Smithers Rapra. "Plastic Design and Material Selection: Risks and Failures."http://www.rapra.net, 2014; LanXess. Engineering Plastics Material Selection: a Design Guide. www.us.lanxess.com.

[7] Albert Einstein, theoretical physicist, 1879–1955. The Saturday Evening Post, *What Life Means to Einstein: An Interview by George Sylvester Viereck*, Start Page 17, Quote Page 117, Column 1, Saturday Evening Post Society, Indianapolis, Indiana, October 26, 1929.

[8] ASQ. "Failure Mode Effects Analysis." http://asq.org/learn-about-quality/process-analysis-tools/overview/fmea.html, 2014; Quality Associate International. "History of FMEA." http://quality-one.com, 2014; see also Fadlovich, Erik. "Performing Failure Mode and Effect Analysis" (Embedded Technology), 2007.

Further Reading

M. Ashby, Materials Selection in Mechanical Design, fourth ed., Butterworth-Heinemann, Burlington, Massachusetts, 2010.

M. Ashby, Materials Selection in Mechanical Design, Elsevier Ltd, USA, 2005, p. 251.

M. Ashby, K. Johnson, Materials and Design, third ed.: the Art and Science of Material Selection in Product Design, Butterworth-Heinemann, 2014.

L. Brown, T. Holme, Chemistry for Engineering Students, Cengage Learning, 2010.

K.G. Budinski, M.K. Budinsky. Engineering Materials: Properties and Selection, ninth ed., Prentice Hall.

E.A. Campo, Selection of Polymeric Materials: How to Select Design Properties from Different Standards, Plastics Design Library, 2008.

E.A. Campo, Polymers in Industrial Applications, Hanser Gardner Publications, 2007.

G.E. Dieter, Overview of the materials selection process, in: ASM Handbook Volume 20: Materials Selection and Design, , 1997.

E.R. Larson, Plastics Guy Blog Entry. "Good Question.", 2013. http://plasticsguy.com/good-question/.

E.R. Larson, Life Cycle Design Guidance Manual, U.S. Environmental Protection Agency, 1993(contributor).

E.R. Larson, "Inside Injection Molded Parts," Materials Engineering, August 1992.

E.R. Larson, Reprinted in Progettista Industriale, in: "Designing Better Thermoplastic Parts," Machine Design, February 1991, July/August 1991.

L.W. McKeen, Effect of Temperature and Other Factors on Plastics and Elastomers, second ed., Plastics Design Library, 2008.

The Macrogalleria Center: A Cyberwonderland of Polymer Fun! http://www.pslc.ws/macrog.htm.

W.J. Strong, T. Yu, Dynamic Models for Structural Plasticity, Springer, 1995.

P. White, L. St Pierre, S. Belletire, Okala Practitioner: Integrating Ecological Design, Okala Team, 2013.

6 Material Selection Based on Cost

Throughout this book, we have stressed that material selection is about finding one or more suitable materials that—in combination with an effective design, proper processing, and eventual integration into a final system—results in a product that satisfies the needs of an end user.

In the previous chapter, we focused on product performance, and discussed how to evaluate thermoplastic materials in order to create a product that meets its intended performance requirements. We have also discussed the concept of price versus performance, where lower cost items have low performance expectations, and higher cost items have high performance expectations. This concept applies not just to products, but also to materials (as well as other things in life). While product performance is an important criterion for material selection, the fact remains that cost is often a critical factor in the commercial success of a product.

In this chapter, we will discuss how to select thermoplastic materials based on cost. For our purposes, when we talk about cost, we are talking not only about the cost of the raw materials, but all of the costs involved in turning those raw materials into a final product. This includes processing costs, handling costs, finishing costs, assembly costs, inventory costs, and delivery costs. While the methodology for cost-based material selection is valid for any kind of product, it is especially useful when designing products that are going to be mass produced. Mass production provides a means for cost-efficient manufacture of parts and products; with effective material selection, these efficiencies can be improved even further, sometimes dramatically.

6.1 What is Cost?

In its most common use, cost refers to the concept of payment. One party provides goods and/or services to another party, and in return that party provides a payment to the first. This payment amount represents the cost of the goods and services to the second party (Figure 6.1).

The terms *cost* and *price* are often used interchangeably. While they are related terms, they are not the same. Price is a number used to establish value. In the case of a transaction involving goods and services, the price may be described as a valuation per a unit amount of goods and services (*the price of milk is $2.19 per gallon*), or as a total amount (*the price of the*

Figure 6.1 A payment for goods and services. sooheekim/Shutterstock.com.

car is $34,215). Price is often a negotiated amount, based on the perceived value of the goods and services, supply and demand, market considerations, etc. Cost is an expenditure, an output of time, energy, capital, or currency. For a given transaction, the total cost of the transaction includes not only the price of the goods and services, but additional costs, perhaps for taxes, duties, and/or shipping and handling.

Prices for raw materials vary, depending on the material, where they are produced, and the markets they are sold in. The markets may be open and efficient, or they may be closed and tightly controlled, either through oligopolies, cartels, or monopolies. Fortunately, the markets for most thermoplastic materials are open, and pricing is stable and globally competitive. However, the cost of the material to the buyer will also include costs for transport (freight, customs, insurance, etc.) and perhaps for warehousing (handling, storage, maintenance, etc.). These costs may be higher than the price of the raw material. The amount of these costs, and the price of the raw material itself, are often defined in sales contracts, which typically specify responsibility for various costs (including duties, shipping and freight, etc.), as well as payment terms. These contracts frequently reference Incoterms (International Commercial Terms) as defined by the International Chamber of Commerce [1].

In thermoplastic material selection, it is important to remember that the cost of goods for the finished part includes more than just the resin price or even the total cost of the material. There are additional costs involved, including processing costs, assembly costs, etc. We will explore some of these cost factors later in this chapter. Most of these costs are obvious and easy to determine, but some are subtle, and others are often hidden (or ignored).

This use of the word *cost* as an expenditure is universal. The units that are used in a given transaction may vary; they may be a currency-based system (local, regional, or global), an exchange of goods in a barter system, or a mix of systems (such as complex transactions involving exchanges of cash, stock, real estate, and other assets). But in all of these transactions, the cost of the goods and services involved is usually measurable

6: MATERIAL SELECTION BASED ON COST

and quantifiable. In many ways, this concept of cost is the fundamental basis for economics.

The term cost is also used to assess the effect of actions or decisions. This may be on a personal level (*What he said was wrong, and it cost him his job*), or on a corporate or even societal level (*The war was won, but at great cost*). In this use of the term, cost can be difficult to quantify, let alone measure. Cost then becomes an ambiguous number with an unknown value or a number that is constantly varying (and being argued about). And while the value (the price) of the effects can be debated, the effects themselves are real. We simply do not have the means to measure and quantify them.

Finally, the term cost is also used to describe the loss of something. While this loss is usually a result of actions or decisions, the term cost is used here slightly differently, with the implicit understanding that what was lost can never be regained. *He tried to save them, but it cost him his life*. In terms of material selection, we often consider the effects of our choices on the health, safety, and welfare of the public. We balance the cost of the materials (the cost of goods) and the cost to society (any loss of health, safety, and/or welfare) and do our best to make effective decisions. It is altogether fitting and proper that we should do this [2].

However, it is rare that engineers and designers are asked to evaluate the effects of their material choices on a personal or societal level. *In terms of climate change, what is the cost of using thermoplastics?* We simply do not have the tools to make that assessment. They involve more than economic analysis, and more than an assessment of health, safety, and welfare. Now one can ignore all that, and simply say, *Hey, it is not my job*. Or one can bury their head in the sand, and leave the job of evaluating the effects to bureaucrats and policy makers, but the fact remains, we are the ones making decisions on what materials are being used, where they are being used, and how they are being used. If we do not make any effort to get informed, we are not doing our job. While, we may not be paid to stay informed about the effects of the materials we use on the world around us, sometimes doing your job involves more than just doing what it takes to get paid. *For what will it profit a man if he gains the whole world and loses his own soul?* [3].

6.2 Why is Cost Important?

When we talk about the cost of a product, we are normally referring to the costs involved in manufacturing and delivering that product to an end user (the traditional *cost of goods and services* definition). While this use of the term seems rather simple, it is an important one.

6.2.1 The Way We Measure Things

In the same way that mass, length, and time are fundamental units of measurement in physics, cost—as in the *cost of goods and services*—is a fundamental unit of measurement in economics. While the units of measurement may vary (based on currency, barter of goods, or exchange of assets), this concept of cost is ingrained in our economic systems. Most of our measurements involving money, or the accumulation of wealth, are predicated on the concept of the cost of goods and services. Even the term inflation is based on this concept.

6.2.2 Relationship to Performance

The cost of goods is also an important aspect of product performance. While product performance can be evaluated in absolute terms (based on a reference to a benchmark), performance is usually evaluated on comparative terms, using a ratio of performance versus cost (or price).

One of the challenges in material selection based on cost is in establishing the proper relationship between performance and cost. What are the expectations for performance at a given cost level? What is the minimum cost in order to achieve a given level of performance? Even with the answers to these questions, product performance is usually the prime focus. The cost of goods is the ugly sister or the jealous brother. And while the cost of goods is important, it is rarely the topic of an advertising campaign.

6.2.3 Business Perspective

In the business world, the basic rules of the manufacturing game are pretty simple: *make something, sell it for more than it costs you to make it,* and *make a profit in the process (aka make money doing it)*. In the context of this game, cost of goods becomes the language of the playing field. Most companies embrace this language and do everything they can to seek competitive advantage by lowering their cost of goods. They negotiate global supply contracts to obtain raw materials at the lowest possible price, they seek out manufacturing partners in various parts of the world with the lowest possible labor rates, and do whatever it takes to manufacture goods at the lower possible cost. While this focus on cost of goods is important, it is often shortsighted. Indeed, many companies have gone bankrupt by focusing their efforts on simply being the lowest cost provider.

6: Material Selection Based on Cost

Another approach to manufacturing is to apply a different strategy, one based not on the lowest possible cost of goods, but on a mix of cost, performance, and timing. In this approach, which I like to describe as *Right Costing*, product performance and delivery are just as important as the cost of goods, and sometimes even more so. In this approach, sales and market share—and profitability—depend on delivering a product with the right level of performance at the right cost at the right time.

6.2.4 The Bottom Line on Cost

On a personal level, we all evaluate the cost of good and services in our daily lives, not just in terms of currency (aka money), but in terms of thought, effort, and life energy. One might even argue that cost is a fundamental unit in the measurement of onc's life energy.

Most of us also equate the term cost with the concept of money. There is an old expression about money that goes something like this: *Money Makes the World Go Around* [4]. And while this is not always true, most of us pay special attention to the cost of goods (Figure 6.2).

There is one unfortunate thing about cost: it is boring. That is right, cost is boring. Talking about cost is boring, estimating costs is boring, reducing costs is boring, even saving money is boring. Sure, while we all like to get a deal on something we buy, nobody is making a musical about cost reduction (see Frank Sinatra in the New Broadway show "Cost Reduction for Manufacturing," featuring the hit song *I got the two penny blues*, and that blockbuster tune, *Manager, can you spare a dime?*).

Figure 6.2 Most of us pay attention to the cost of goods.

So perhaps, instead of describing this chapter as Material Selection Based on Cost, perhaps we should give it a different title? Something hip, edgy, catchy. Something like one of the following:

Material Selection Based on Making Money

Material Selection Based on Accumulating Wealth

Material Selection Based on a Comprehensive Methodology for the Assessment of Personal Satisfaction and Societal Impact

On the other hand, perhaps we will just stick with Material Selection Based on Cost.

6.3 The Language of Cost

Another major challenge in selecting materials based on cost has to do with language. Cost is a complex subject and the word itself has a number of meanings. Even if we focus on the concept of cost as the payment amount in a given transaction, there can be a number of ways that the payment amount can be measured (and argued over).

We also need to understand the difference between cost, as a payment amount in a given financial transaction, and accounting, which is the measurement and analysis of a number of financial transactions.

Accounting is often described as being the *Language of Business*. This may be a bit of overstatement, because there are many different types of accounting, just like there are a many different languages. There is management accounting, tax accounting, financial accounting, lean accounting, throughput accounting, forensic accounting, social accounting, and many more.

One of the most common types of accounting, especially in regards to a manufacturing business, is the practice of financial accounting, which is used to evaluate a number of variables of a business, including its assets and liabilities, its business value, its income and expenses over a certain period of time, etc. While material selection can have an effect on the profitability of a business, financial accounting practices are not very useful in the process of material selection.

A much more useful tool for material selection is the practice of cost accounting. Cost accounting is a form of management accounting and is often used to compute the cost of goods, services, projects, processes, etc., in order to assist management in their decision-making. In this type of accounting, the focus is on tracking, recording, and analyzing all of the financial transactions involved in the delivery of goods and services, and

6: MATERIAL SELECTION BASED ON COST

there are usually very specific ways of measuring the value of a given transaction. Unfortunately, most cost accounting systems are not governed by rules and standards such as the generally accepted accounting principles used in traditional financial accounting. Cost accounting methods vary from company to company and sometimes even within a company. (*You like to-may-toes, I like to-mah-toes* [5].)

To help guide us in the process of material selection, I propose the use of a number of generally accepted cost principles. They are listed below. While the method of measurement for each of these may vary from company to company, they can still provide a design team with useful cost data, provided the measurements are done in a consistent manner.

Capital—the amount of wealth, in either money or assets, that is owned by an individual or company.

Capital asset—property that is owned by an individual or company. The property may be physical (such as land, buildings, and equipment) or virtual (such as a trade route, or trademarks, patents, or other intellectual property).

Capital investment—money that is invested in a company or capital assets that are acquired.

Amortization—the process of allocating the cost of a capital asset over an extended period of time.

Cost of goods—the total cost involved in manufacturing and delivering a part and/or product to an end user.

Material cost—the cost of the raw material used in a part or product.

Processing cost—the cost involved in converting raw materials into a functional part or product. Processing cost can describe the cost of a specific manufacturing process or the sum total of a number of processes.

Tooling—capital assets that are procured to manufacture a specific part or assembly. Tooling is normally only useful for the part or assembly that it was procured for, and may involve molds, dies, forms, cutters, fixtures, etc. Tooling is an important aspect of plastics manufacturing, and tooling costs are affected by part design and material selection.

Injection mold—a specific kind of tooling used to fabricate parts in the injection molding process. Injection molds are typically the most expensive kinds of molds used in plastics processing.

Tooling maintenance—costs that are incurred in the care and upkeep of tooling to ensure it is suitable for production purposes. Tooling maintenance may be included in the original tooling cost, allocated to the process cost, or treated as a unique cost expense.

Tooling ownership—Unlike other equipment used in plastics processing, tooling is normally procured to manufacture a specific part or product. Often, the purchaser of the part or product will pay a fee up front to cover of cost of tooling. It is often assumed that the purchaser "owns" the tooling. However, without written documentation to support that claim, ownership of the tooling cannot be certain.

Boat Anchor—a slang term, used to describe an injection mold that is no longer suitable for fabricating parts. It may be because the design is obsolete or the mold is not capable of producing satisfactory parts. Note: being associated with a boat anchor is not good for one's career.

Design—an outline, sketch, or plan to create something; also the activities involved. In manufacturing, good design decisions can decrease the cost of goods and increase the wealth of a company. Conversely, bad design decisions can increase the cost of goods and decrease the wealth of a company (and sometimes result in boat anchors).

6.4 Evaluating Cost

Evaluating the costs of goods is a fundamental business exercise. It is a task that any first year accounting student can complete. Unfortunately, many engineers and designers do not know how to do it. They develop a design, select a material based on an evaluation of some property data, and then send out a request for quote (RFQ) for the parts. Then, when the quotes come back, they are stunned that things cost so much.

Some development teams take a more proactive approach to part cost. As part of their product definition they may even create a set of cost specifications for the final product with specific cost targets for each component. A short list of candidate materials is then prepared and cost data (resin pricing) for these materials are collected. The design team then reviews the data, compares them to the defined cost specifications, and selects a few specific materials for further review. As the project develops, the design team refines the design and continues with their evaluation of cost using the selected materials. While this method is better than the standard *RFQ and Hope* approach, it still may overlook many important cost factors.

6: Material Selection Based on Cost

At their most fundamental level, part costs are a classic example of the cost of goods and services. There are two major components to part cost, the first being the material cost (the cost of goods) and the second being the processing costs (the costs of services). The basic equation is as follows:

Part Cost = material costs + processing costs

The overall equation is simple, and while at first glance it seems that the two cost items are separate entities, material selection affects both costs items, although in different ways. We will explore some of these effects a little later in this chapter. Also, it should be noted that our use of this equation is intended to reflect the total cost per part on a prorated basis. This allows us to account for costs for yield and scrap, downtime, overhead, storage, etc. While these costs can be broken out and tracked individually (either on a part basis, or a process basis, or at a factory level), for our purposes we will simply integrate them into the total part cost.

6.4.1 Material Cost

Let us look at the first half of the part cost equation, which is used to calculate material costs.

Part Cost = **material costs** + processing costs

where

$$\frac{\text{material costs}}{\text{part}} = \frac{\text{material mass}}{\text{part}} \times \frac{\text{material price}}{\text{unit mass}}$$

and

material mass = f (design)
material price = f (resin pricing, supply chain, etc.)

The material cost equation is also quite simple, and there are only two primary variables: the mass of the part, and the material price. (Again, these variables are prorated numbers that account for all of the material used in the production of the parts in question, including scrap, setup, and changeover, etc.)

In order to reduce material costs, many companies focus their efforts on minimizing the material price. This is an important exercise, especially in high-volume production where the overall quantity of material being used is substantial. Reducing the resin price by a few pennies per pound can add up to significant cost savings. However, the mass of the part plays a significant role as well, and the mass will depend on the design of part. One can have the world's greatest negotiator finalize the price on a global supply contract for a given resin, but if the design team has not done their homework the overall material costs are still going to be higher than they need to be.

6.4.2 Processing Cost

Now let us look at the second half of the part cost equation, which is used to calculate processing costs.

$$\text{Part Cost} = \text{material costs} + \textbf{processing costs}$$

where

$$\text{total processing cost} = \text{molding costs} + \text{handling costs} + \text{finishing costs} + \text{assembly costs} + \text{packaging costs} + \text{inventory costs, etc}.$$

and

$$\text{process}_n \text{ cost} = \text{process}_n \text{ rate} \times \text{cycle time}_n$$

$$\text{process}_n \text{ rate} = \frac{\text{cost of process}_n}{\text{unit of time}}$$

The process cost equation is also quite simple, and while there may be many processes involved, there are only two variables in each process: the process rate (the cost of the process per unit of time) and the cycle time of the process (how long it takes). Let us explore these two variables in greater detail.

6.4.3 Process Rates

In economics, the term *rate* is used to describe a specific kind of financial transaction that is based on proportions. The term has its origins in the Latin phrase *pro rata*, which roughly translates to "according to a calculated amount." Some common examples of financial transactions that are based on rate include taxes, interest, commissions, hourly wages, etc. Even a salary is a rate-based transaction (with your paycheck being based on a proportional amount of your salary per month or per year).

In the world of manufacturing, there are a number of rate-based transactions that affect the cost of goods. These include labor rates, machine rates, and overhead rate, which is a proportional number based on an allocation of the fixed and variable costs that are incurred in operating a factory. A process rate is the total of the labor rates, machine rates, and overhead rates for all of the labor and machinery involved in a given process.

6: Material Selection Based on Cost

As an example, for a molding process, it involves the machine rate of the molding machine itself, plus the labor rate of the operator (assuming there is one), plus the overhead rate of the factory. These rates can be calculated a number of different ways, depending on how asset costs are amortized, how overhead is allocated, how maintenance and repairs are accounted for, etc. Labor rates will vary, depending on the skills involved, and the region where the factory is located. Machine rates will vary based on the size and complexity of the equipment, as well as its age.

Determining the exact rate for a given process can be a complicated accounting exercise, but there are often aspects of that rate that can be easily estimated. Furthermore, it is often helpful to adjust these estimates into numbers that are easy to work with. For instance, there are 60 s in a minute and 60 min in an hour. So, there are 3600 s in an hour. If you have a rate of $35.22 per hour, round it up to $36 per hour, which equates to $0.01 per second—a penny per second. This is an easy number to remember and an easy number to work with. And while a penny per second may not seem like that critical of a number, the seconds mean pennies, pennies turn into dollars, and the dollars add up.

Below are some tables which can be useful for estimating costs. The first table gives values for various units of time, based on a work schedule of 8 h per day and 5 days per week (with no holidays or breaks) (Table 6.1). The second table gives some values for various rates, with some conversions to numbers that are easy to remember (Table 6.2). These numbers are also based on 8 h per work day and 5 work days per week (with no holidays or breaks).

By the way, you may have heard the phrase, *a penny for your thoughts* [6]. If you are being paid $36.00 per hour (which is just under $75,000 per year), in order to justify that penny for your thoughts, you would need to come up with a new idea every second of every working day—otherwise, you are being overpaid.

6.4.4 Cycle Times

In manufacturing, the phrase cycle time is used to describe the amount of time (as measured in seconds, minutes, hours, or days) that is required to complete a given manufacturing process. In the case of thermoplastics, most processes have cycle times that can be measured in seconds or minutes. In some cases, multiple parts can even be processed during the same cycle. When this occurs, the cycle time is prorated to a per part basis. As an example, if a molding process takes 60 s, and four parts are molded at once, the per part cycle time is prorated to 15 s.

Table 6.1 A Time Table

Units of Time				
Work Weeks	Work Days	Hours	Minutes	Seconds
	0.002	0.017	1	60
0.025	0.125	1	60	3600
0.20	1	8	480	28,800
1	5	40	2400	144,000
4	20	160	9600	576,000
10	50	400	24,000	1,440,000
52	260	2080	124,800	7,488,000

Table 6.2 A Table of Cost Rates

Rates (Currency = US $)						
Per Year	Per Month	Per Week	Per Day	Per Hour	Per Minute	Per Second
$260	$22	$5	$1.00	$0.125	$0.002	$0.00003
$1248	$104	$24	$4.80	$0.60	$0.010	$0.00017
$2080	$173	$40	$8.00	$1.00	$0.017	$0.00028
$2600	$217	$50	$10.00	$1.25	$0.021	$0.00035
$7488	$624	$144	$28.80	$3.60	$0.060	$0.0010
$12,480	$1040	$240	$48.00	$6.00	$0.100	$0.0017
$20,800	$1733	$400	$80.00	$10.00	$0.167	$0.003
$74,880	$6240	$1440	$288.00	$36.00	$0.600	$0.010
$2,08,000	$17,333	$4000	$800.00	$100.00	$1.667	$0.028

Cycle times are an import criteria in part costs, and molders and part suppliers pay close attention to them. Unfortunately, cycle times are rarely addressed in the design phase and as a result they are ignored in the initial material selection efforts. While molders may seek to influence the selection process by recommending additives or material grades with superior processing characteristics (and rightly so) quite often the design team shrugs this off as a manufacturing issue and does not appreciate that it is also a cost issue.

In addition, cycle time data for the processing of thermoplastic materials are rarely provided by resin suppliers—regardless of the process. Cycle times are going to vary depending on the material, the size of the part, the thickness of the part, and the process used. But even with all of the design tools that exist, cycle time estimates are usually based upon first hand experience, and actual cycle times are often a closely guarded secret (sometimes even classified as proprietary and/or confidential information). While the cycle time data itself may be difficult to obtain, their use in the basic cost equation is straightforward and easy to track.

6.4.5 Adding Up the Numbers

Going back to our original cost equation, we now have a modified equation that goes something like this:

$$\text{Part Cost} = \left(\frac{\text{material mass}}{\text{per part}} \times \frac{\text{material price}}{\text{per unit mass}} \right) + \sum (\text{process}_n \text{ rate} \times \text{cycle time}_n)$$

To determine the total part cost, we simply add up all the numbers, and if we want to reduce the cost, we reduce the value of one or more of the input variables. Determining the optimal cost can be a mathematical challenge, but there are some fundamental things that can be done to reduce part cost on very basic level, some of which we will discuss shortly.

What is not immediately obvious is the fact that material selection affects every single variable in this equation. If you are selecting materials without considering these effects, you are not doing your job.

6.5 Reducing Material Costs

To reduce material costs, many manufacturers focus on resin price and they take extraordinary steps to purchase resin at the lowest possible price. They will hammer on their suppliers to keep resin prices down, negotiate to lock in global supply contracts, and sometime pit suppliers against each other through practices like blind auctions. At first glance this makes sense, because the lower the resin price, the lower the raw material cost.

However, there are often other factors that affect the overall cost of the raw material. These may include handling and inventory management costs, shipping and/or freight costs, costs for additional Stock Keeping Units (SKUs) for different materials, accounting costs, etc. Many of these items fall into the category of supply chain management. While cost data for many of these variables may not be available to the design team, it is important to understand that they exist (and having a supply chain professional as part of the design team can be a huge asset [7]).

Even with a comprehensive plan for supply chain management there are still many opportunities to reduce material costs. As we stated earlier, material selection affects every single variable in the part cost equation. Since design and material selections are interrelated, design also affects every single variable in the part cost equation.

It is the responsibility of the design team to create a design that achieves the right level of performance at the right amount of cost. This is not a sourcing task or a supply chain task, it is a design task. In this section, we will discuss some of the things the design team can do to reduce part costs, specifically the material costs. While these items are presented in a sequence, when and where they are used in the design process may vary, depending on the application. In some applications, they can be used in this sequence, other applications may call for a different sequence. On rare occasions, some items can be done concurrently or even skipped.

6.5.1 Optimize the Structure

In every application, there is a primary structure. This structure forms the backbone of the part. It usually provides the required strength and stiffness, and it may also provide a foundation for other part features. The first step in reducing material cost is to determine what material will provide the most cost-effective structure. The selection may involve other aspects of performance, such as toughness, and considerations for the end-use environment (exposure to heat, chemicals, etc.), but it should also include a calculation of the material cost that is required for the structure.

Sometimes the cost is easy to calculate, such as for a simple beam in a specified space with specified flexural and torsional requirements. As you design the beam for the required stiffness, you can make a simple spreadsheet with dimensional data and use it to track deflection. You can use the same data to calculate the part volume and then calculate material cost per part for that material in that design. You may find that you can design a beam in a glass-reinforced nylon that is cheaper than any beam you could design using talc-filled propylene, even though glass-reinforced nylon is a more expensive material per pound. This is exactly the case with the classic five-legged bases that are used on office chairs (Figure 6.3).

On other projects you may have a certain amount of design freedom. Perhaps, the performance requirements are not very stringent, and you

Figure 6.3 Office chair with five-legged base. DmitryKolmakov/Shutterstock.com.

realize that there are several different materials that could be used, provided you modify the design slightly. Depending on which material is used, you might need to add or remove features, use more (or fewer) ribs, or use a different kind of assembly method. One material might require you to use threaded fasteners (with or without inserts), while with another you can make use of a snap fit. Consequently, each of these designs will be unique, and each will have a different material cost.

For each design, you can then analyze the structural performance, determine the amount of material that is required, and then calculate the resulting cost of that material. These calculations are quite simple with standard CAD software. One can use different CAD models for different design approaches, and track the part volume and material cost per part for each concept as the designs evolve.

Very quickly, one can usually determine which material will provide the most cost-effective structure. Once that is complete, you can then shift design efforts to refine and optimize the structure using that material, making it even more cost-effective.

It is important to remember that you are trying to reduce the cost of the material per part for a specific material in a specific design. You are not just swapping one material in and out of the same design and comparing resin prices. Your calculations must account for the design changes involved when using different materials.

6.5.2 Effective Specifications

Just as we can refine the structure to optimize the amount of material that is required (and its cost), we can also refine the specifications to further optimize the cost of the material being used. This refinement involves product specifications and material specifications.

A product specification should describe *what* the product must do (not *how* it should be done). Usually the product specification is written before the design process begins, but sometimes, it is written concurrently with the design process. Regardless of when it is written, the performance requirements of the product should be clearly articulated. As we have discussed previously, performance and cost are interrelated and the performance requirements will have cost implications for every aspect of the design.

In terms of individual parts, the materials that are selected will typically be described in a material specification, either on a part drawing, or in a material specification document. This material specification will describe the chemical family, the type (homopolymer, copolymer, etc.), the additives, perhaps the color (or lack thereof), and more. It may even reference some standard property values of the material and specific product codes from approved suppliers.

What is sometimes overlooked in a material specification is the availability of the material. We all have stories about how we were trying to buy something simple, only to find that this certain something—from that company, in that color, in that size, on that day in that city in that whatever form it was that we wanted—involved a special order. Not only was that order going to take a few extra days, it was a *Special Order*, so it was going to cost more. *ARRRGGG*! This exercise also applies to materials, especially thermoplastic materials. The cost of a given material is often directly related to its availability.

As a general guideline, one should select and specify materials that are readily available in the region where the parts will be manufactured. While this may seem like common sense, sometimes there are unforeseen issues. Perhaps, the product that you designed was originally intended to be manufactured in China, but due to global demand, your company is now setting up facilities in Belgium, Brazil, South Africa, and Australia. You had selected a standard material in a standard grade in a standard color that—according to the Web site you looked at—was available in the

6: MATERIAL SELECTION BASED ON COST

Asia–Pacific region (and, the last time you checked, China was part of the Asia–Pacific region). However, you have just learned that the "standard" color you specified is not exactly standard, it is simply a color that was created for another application, which was then given a "standard" commercial code which was then added to the supplier catalog. To make matters worse, not only is that material in that grade in that color not readily available anywhere in China, that color code is not available anywhere else in the world (unless you would like to make a *Special Order*). On top of all this, your boss has scheduled a meeting with you to discuss your future with the company.

Specifications also often overlook the material grade. By default, the standard assumption is that the material must be a prime, 100% virgin grade, with no regrind allowed. While there are many industries where this is a requirement, there are also industries and/or applications where a lesser grade can be equally effective, for example, a generic prime grade, or an industrial or off-spec grade, or perhaps, even a reprocessed or recycled resin. If your material specification addresses the use of these grades, it can allow for the use of a substantially lower cost material—but one that performs equally as well.

In a similar manner, there may even be an opportunity to make the material specification as broad and generic as possible. Instead of defining a single unique material, the specification can allow for the use of multiple materials: perhaps, one material is a general purpose Acrylonitrile butadiene styrene (ABS), and another is a talc-filled polypropylene. Assuming both materials work equally well in the application, by writing a more generic specification you cannot only allow for the use of different resins from different suppliers, you can also allow your sourcing team to take advantage of periodic fluctuations in resin pricing by using an entirely different material family. While applications like this are uncommon, when they occur, take advantage of them, as they can provide substantial cost savings.

One unfortunate thing about material specifications: they are boring. Most of them—even well written specifications—provide nothing more than a list of approved suppliers, some commercial product codes, and perhaps some basic material properties. They are the industrial equivalent of asking, *Could I please have some water?* (and then failing to mention, *Oh yeah, I would like the water to be potable, free of contamination, and served at room temperature, in a 12 oz drinking glass with straight sides and no flutes without any ice or lemon slices. And no substitutions without my approval, thank you.*)

In order for a material specification to be effective, it should reference an approved technical standard, one established by a trade organization, or a governmental or regulatory agency. Among other things, these types of standards will ensure consistency, reliability, and traceability of the source material.

However, in order to be cost-effective, a material specification should also address the commercial availability of the material (either as standard or special order), the required grade of the material, and what deviations from the standard are acceptable (if any). While providing this level of detail may involve more work, it can allow for a substantial reduction in material costs.

6.5.3 Optimize the Wall Thickness

Next, one should design parts to use the minimum amount of material. This means not only creating a design with the most cost-effective structure, but also by optimizing the wall thickness throughout that structure. The wall thickness should be thick enough to account for structural needs, and for processing needs, such as the material flow and fill during the injection molding process. It should also be thick enough to account for the required shape and structure needed for any subsequent manufacturing steps, including secondary processes, assembly methods, and shipping and handling. However, the wall thickness should also be as thin as possible in order to reduce the amount of material that is used. Remember, the less material that is used per part, the lower the material cost will be.

6.5.4 Exploit the Material

An experienced plastics engineer should also explore ways to exploit the nuances between different materials. As an example, you may find that a certain material flows better, which can allow for a thinner nominal wall thickness, or require fewer ribs for flow and fill (either of which will reduce overall part volume). Or you may find that an impact-modified material can allow for smaller corner radii (which can also reduce part volume). Perhaps, a material you are considering can also allow for unique features, which will improve the structure or the assembly (or both).

Several years ago, I was invited to participate in a design project for the BIC Corporation. BIC is a global manufacturer of high-quality, affordable consumer products. This project involved a possible design change to one of their products, the BIC disposable lighter. This small, inexpensive device provides users a simple means to make fire on demand—at the flick of a BIC. It does this safely, reliably, and cost effectively—exactly as it was designed—for millions and millions of people around the world. It is a design and engineering masterpiece included in the architecture and design collection of the New York Museum of Modern Art.

BIC has produced over 20 billion lighters in the last 35 years. Worldwide, they produce over 5 million lighters every day. The main body is made from acetal homopolymer. The main body functions as a reservoir for the butane gas which fuels the flame. Acetal is a great engineering material—it is stiff, strong, has good impact resistance, and great chemical resistance as well.

Acetal also has a combination of properties that are ideal for snap-fit assembly. It is reasonably hard, has high lubricity, and also allows for molding of very fine part details. The BIC engineering team had taken advantage of these properties and had designed a set of support features into the top of the reservoir. These supports held the flint wheel and release lever in place, acted as a pivot point for the release lever (also molded out of acetal homopolymer), and provided mating features for the metal flame guard. All of these features worked together to allow the lighter to function easily and almost effortlessly, and they all snapped together without need for fasteners or adhesives. The overall design was simple, used a minimal number of parts (lower material costs), and allowed for lightning fast assembly (lower processing costs). There are very few thermoplastic materials that could meet all of the defined performance requirements, and after selecting acetal (based on the fact that it provided the most cost-effective structure) the BIC team successfully exploited the nuances of the material to lower their overall costs (Figure 6.4).

Another example of exploiting material nuances comes from the sporting goods industry. Underwater Kinetics, a San Diego manufacturer of

Figure 6.4 BIC lighter and parts.

dive lights, had started to expand their business focus, and was developing products for fire and rescue, industrial safety, and general sports activities. I was responsible for bringing a new LED headlamp design into production. Code named Vizion, it was a small device, but with lots of features, including a pivoting body, an adjustable beam, and multiple lenses. It was also waterproof to a depth of 30 ft (10 m).

The lens of this headlamp was molded from polycarbonate (stiff, strong, impact resistant, and has excellent light transmission in its clear form). The lens itself was a Fresnel type lens and the part also had some molded-in mounting pins. We had a prototype mold fabricated to prove out the optics and when the first parts were tested we noticed some problems. The optics of the lens was fine, but the mounting pins were not quite strong enough. So we started exploring ways to improve the structure.

As we discussed ideas, we also reviewed issues we had with other parts, and with overall fabrication and assembly. Each idea led to another and we soon had a completely new design for the lens. It retained the Fresnel lens for optics, but we replaced the pins with two large mounting bosses that provided additional structure to keep the lens flat and dimensionally precise, along with an interface that would also act as a locating datum. We also added a perimeter flange that interconnected all the features and providing a protective shroud for the electronics. By texturing this flange, we created a visual barrier, one that also acted as a diffuser for low-level lighting needs. We were able exploit some of the unique characteristics of polycarbonate to not only provide the needed structure, but add needed functionality and reduce the overall part count (and overall material cost) (Figure 6.5).

Figure 6.5 Fresnel lens with mounting bosses and shroud.

6.5.5 Exploit Competitive Advantages

Finally, whenever possible, one should design parts that exploit competitive advantages in the supply chain. As a simple example, if your client or employer is a high volume user of a specific material, they will usually have a supply contract with a supplier who provides them with that material at a favorable price. If you can use that material in a specific part to gain a cost advantage—on either a regional or global basis—by all means do it. It might be an over performing material for the application, but so what?

Just like buying in bulk a discount warehouse, a high volume user of a material will almost always get more favorable pricing than low- to mid-volume user. While this favorable pricing may apply to all of the materials that are being purchased, usually it only applies to the resin being purchased in bulk. However, it may sometimes be an advantage to consolidate the resins that are being used. This can allow for a better price on the high-volume material and can also reduce inventory costs.

In the early 1990s, the SRAM Corporation began selling a bicycle shifting mechanism under the brand name Grip Shift (Figure 6.6). It consisted of a rotating plastic cylinder that acted as a cam to pull and release the shifting cable in a precise manner. It was a very simple

Figure 6.6 SRAM IBS shifters.

solution to a complex problem. The device sold well, and they soon began making different versions of the shifter, targeted at different price points and performance levels. As their sales grew, they also began making other bicycle components, including derailleurs, brakes, and brake levers. All of these products made extensive use of thermoplastics. At one point they were using something like eight different thermoplastic materials, sourced from several different suppliers. Some of these materials were purchased in very low volumes and the pricing for each material was substantially different. We began to look at ways to reduce our material costs.

One of the opportunities we pursued was a strategy of material consolidation. Instead of using many different materials, we focused on using a few key materials, using a concept of good, better, best. For basic applications, we specified a general purpose ABS. For higher performance applications, we specified a 33% glass-reinforced 6/6 nylon. And for ultra high-performance applications, we specified a toughened, glass-reinforced aromatic polyamide. We went from eight different materials, to three standard materials. The production volumes of each material went up and our costs went down. In addition, the entire team—from the design staff to the manufacturing floor—developed a better understanding of the properties and behavior of each material.

Competitive advantages in the supply chain often extend well beyond favorable pricing. Most suppliers, regardless of whether they supply materials, equipment, or services, commit substantial resources to support the needs of their customers. These resources may involve help desks, dedicated account managers, or access to technical staff with expertise in materials, processing, or advanced technology. Effective use of these resources can allow you to reduce your material costs, and may also help in the development of new products.

One of my favorite examples of the using resources in the supply chain came early in my career when I was working for Kransco Manufacturing. Kransco made number of recreational products, mostly in the swimming and surfing area. One of their most well-known products was a line of body boards, sold under the brand name Morey Boogie.

A body board is a short surfboard, usually made of foam, which allows the user to lay on top of it, and then ride the surf in a manner similar to body surfing. It is easier to learn than traditional surfing, which involves standing up on a larger, rigid surfboard. Many surfers look down on body board users, calling them boogers or spongers, although body boarding can be an elegant and technically advanced skill all its own.

Body boards are made from several layers of closed cell polyethylene foam. The main component, called the core, was made from a solid plank of foam, which was then cut to shape, and then covered with thin sheets of foam known as the skins. The performance of the board was determined by density of the core, the thickness and number of layers in the skins, and the overall shape of the board. A primary supplier of the core material was the Dow Chemical Company, under the brand name Ethafoam®.

Kransco was a high volume buyer of Ethafoam®, and received favorable pricing and excellent customer support. My boss, the Director of Product Development, had previously worked for Dow. As a result, Kransco was often the benefit of developments in polyethylene foam technology. One such development involved laminating a thin film of Surlyn® to a sheet of Ethafoam®.

Surlyn® is a brand name of a material that was developed by the DuPont Company. It is an ionomer of polyethylene, meaning the polymer chain had been modified via ionization. It is a flexible and highly durable material, and is also smooth and slippery, with a low coefficient of friction. One of its early uses was in golf ball covers, where it is still used today. By laminating a film of Surlyn® to a sheet of Ethafoam®, one could have a roll of material that could be easily transported to its end-use location, and then cut into shape, and installed in place. One interesting use was in horse stalls, where sheets could be installed on the floors and walls. It was durable enough to withstand day-to-day use, protecting not only the horse stall but the horses themselves. The slipperiness also meant horse manure would not stick to it, allowing the stalls to be easily cleaned.

Thanks to our connections in the supply chain, we obtained a sample sheet of this material, which was dark brown in color (I wonder why?). We were able to laminate this sheet to the bottom of a boogie board, and sent some prototype boards to our test crew, which included Jack Lindhom, and some of the other top professional body boarders in Hawaii. The response was overwhelmingly positive. The new board was faster, more maneuverable, and generated excitement wherever it was shown. We quickly developed a new commercial design which was named the Mach 7-7. Thirty years later, the Mach 7 product line is still in production (although the Surlyn® film layer is no longer brown) (Figure 6.7).

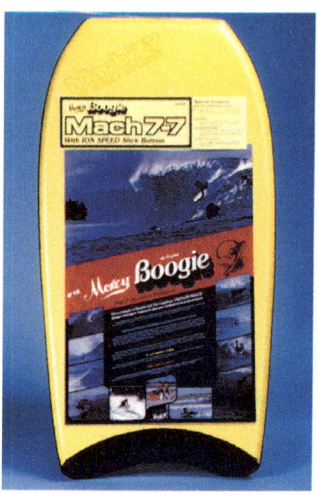

Figure 6.7 Mach 7-7 boogie board, made of polyethylene foam with a Surlyn® bottom skin.

6.5.6 Saving Pennies

The ideas we have discussed for reducing material costs are fairly basic. They represent a number of tried and true methods, and while they may not always result in cost reductions, one of them might help you save a few pennies here and there. In the world of mass production, pennies turn into dollars, and dollars add up. Furthermore, it is important to remember that material costs are usually direct, up-front costs, and a reduction in material cost can create an immediate improvement in the profitability of the business.

6.6 Reducing Processing Costs

The second half of the part cost equation is used to calculate processing costs. This includes not only the cost to mold or form the material, but all of the processes that are involved in taking the material from its initial raw state and turning it into a finished product. Many of these costs are rarely considered by the design team.

Fortunately, there are a number of initiatives in the manufacturing industry that are focused on processing costs. These initiatives are usually described in terms of D F X, where the letters D and F refer to the words *Design For*, and the letter X refers to the specific goal of the initiative. Some of these initiatives include DFA (Design for Assembly), DFM

6: Material Selection Based on Cost   231

(Design for Manufacturability), DF6σ (Design for Six Sigma), etc. These initiatives have resulted in a number of excellent design guidelines, and by incorporating some basic cost data on process rates and cycle times, a skilled design team can use them to achieve great results.

In regards to the use of thermoplastics, there are also a number of design exercises that can be undertaken to reduce costs even further. A few of these are discussed in this section.

6.6.1 Optimize the Geometry

Processing costs for parts made from thermoplastic materials are usually significantly lower than for parts made from metals. Plastic parts can often be fabricated into a final shape with fewer process steps and the cycle times for these processes are often faster as well. Injection molding in particular can usually provide a near net shape part faster than almost any other manufacturing process, and the per part processing cost is very low. With subtle adjustments to the part geometry, the processing cost for a given thermoplastic part can often be reduced even further. These adjustments can facilitate lower cycle times and sometimes even lower process rates.

We need to remember that part geometry is the fundamental driver of processing cost. Part geometry includes not just the part size and shape, but the number of features, and the complexity of those features. In the case of injection molding, the part geometry will dictate the design of the mold, which in turn often dictates the machine it runs on, and the cost of running that machine. This is true not just for injection molding, but for almost all molding and forming processes, as well as handling and assembly.

When designing parts and products made from thermoplastics, we often think of design freedom. Plastic can allow for advanced forms, curves and slopes, and complex 3-D surfacing, in ways that other materials cannot. Yet, we often do not consider what implications these advanced forms may have on the molding and forming processes, or on subsequent processes.

While design is a 3-D exercise, most manufacturing processes are based on 2-D technology. Yes, there are multiaxis machining centers and multipurpose programmable robots, but most equipment still operates in a 2-D manner. Milling and drilling is often up and down, in and out. Most molds open and close in a linear manner as well. And while they may have moving side actions or retractable cores, these motions usually occur in a linear manner. The challenge in the design process is to balance your 3-D forms with 2-D equipment constraints.

As an example, it is a common practice for most designers of plastic parts to envision the layout of an injection mold, based on a certain part orientation. They will use this orientation to establish a parting line, and then add specific features to be on either the fixed or moving half of the mold, along with draft and radii. Quite often the function of the part will dictate the orientation of the part in the mold. It may need to have a flat interface at assembly (which also makes for an ideal parting line) or it may have cosmetic surfaces that cannot have a parting line, etc.

An important thing to remember is that the mold layout usually determines the overall size of the mold and in many cases the size of the mold determines the size of the molding machine. If these choices are intentional, that is fine, but if they are made by default, or done without thinking, you may force yourself into using a machine that is larger—and more expensive to run—than you really need. This is particularly important with multicavity tools.

As a suggestion, as you envision the orientation of the part (or parts) in the mold, explore different ways to orient them. Start by evaluating the footprint of each part. This is the cross section at the parting line. From there, you can determine the cross-sectional area, which determines the required clamping pressure, and when multiplied by the number of cavities, will give you the tonnage required for the molding machine. Different orientations of the part will result in different footprints and different mold layouts. Try to select an orientation that provides the optimal combination of simplicity and size. Ideally, you would like to run each mold on as small and as inexpensive a machine as possible, while still providing the desired level of performance for each part—and for the end product.

There are times when you may choose an orientation of a part based on other criteria. A certain orientation may provide for more robust molding, or better mold reliability, or more accuracy in the molded parts. So even though the mold itself ends up being larger, and may need to run on a larger, more expensive machine, it can still provide for lower overall processing costs in the long run. Keep in mind this general rule: the simpler the mold, the more reliable it will be, and the faster the cycle time.

Also keep in mind that the layout of the mold can have an effect on dimensional tolerances. Dimensions on features that are exclusively in one half of the tool (known as cavity bound dimensions) will usually be more precise than dimensions on features that span parting lines, and are formed by multiple parts of the tool. Dimensions that are aligned in the flow direction may also have different precision than dimensions that are perpendicular to the flow.

This same approach applies to the design of jigs and fixtures that are used in assembly. Again, the idea is to adapt your geometry to accommodate the constraints of the equipment that is involved, in order to make each process as time efficient as possible.

Some companies try to standardize this kind of design optimization into a D F X system, such as DFA or DFM. Other companies segregate the product design team from the mold design team and both of these teams from the manufacturing team. As a result, there may be many wonderful acronyms and guidelines based on a variety of D F X philosophies, but a comprehensive design plan is missing. I prefer to utilize a more comprehensive design approach, what I like to call Design for 42 or DF42 (42 being the answer to life, the universe, and everything [8]). The intent is to enable the design team to Design for All of the Processes involved.

Regardless of what you may call the overall approach, it is advisable to follow standard practices. (They exist for a reason.) These practices are robust, reliable, and repeatable, and as a result they are usually time efficient and cost-effective. Pay attention to the numbers—machine sizes, process rates, cycle times, etc., and continue to refine the design to optimize the process costs.

Another standard practice is for a manufacturing team to optimize the number of molds and/or the number of cavities per mold based on an estimate of molding cycles, production requirements, and machine availability. If you take this one step further, you can add in data for machine rates, and further optimize the number of molds and/or cavities per mold based on the cost to mold the parts in different groupings of molds/cavities/machines. While you may think of this exercise as something for the manufacturing team to deal with, part geometry will be a driving factor, and the optimization process will be influenced by design.

6.6.2 Effective Specifications

In earlier sections, we discussed product specifications, and the relationships between product performance and product cost. By creating more effective specifications, we can ensure the cost of the materials used in our products. In a similar fashion, there are also times where we can create specifications, even high-performance specifications, that can enable a reduction in processing costs, allowing for a design solution with a lower total cost.

Product specifications typically describe the function and performance of a product. Part specifications on the other hand, very rarely describe the part function. Rather they specify what the part should be made of, and it's exact weight, shape, and size (and sometimes even how all of these things are to be measured). The function of the part—its purpose or reason for existence—is often described in a higher level specification.

Collectively, these specifications are used throughout the design and manufacturing process. One purpose is to control the form and fit of the component parts so that we can guarantee that everything always fits together, and the parts and the product always function as intended.

In the engineering world, the word *tolerance* is used to describe the allowable variation in a measured value. When it is used in association with thermoplastics, or manufacturing in general, it usually applies to the measurement of physical dimensions, mass, length, etc. (It may also apply to molecular weight distribution or melt flow measurements, etc.) However, another meaning of the word *tolerance* has to do with acceptance of things that are different. We can rely on both of these meanings to provide insight into reducing processing costs.

Most designers and engineers are familiar with dimensional tolerances. Indeed, tolerance analysis of part dimensions is a key engineering activity; one that helps ensure the fit and function of all of the parts in a final product.

Unfortunately, it is often difficult to determine the appropriate tolerance value for a given dimension on a given part. Production tolerances are typically a function of a number of variables, including the type of dimension (mass, length, position, etc.), the magnitude of the dimension, the material used, the manufacturing process, and the equipment used in that process (as well as the skill and expertise of the provider of the process). The tighter we make the production tolerances (that is the higher our requirement for the precision of the part), the higher the processing costs are going to be. To improve precision, we need newer, better machines (which are almost always more expensive), more precise monitoring of process variables, more frequent maintenance and calibration, perhaps even operators with specialized expertise (who are usually paid more). All of these items add cost. Precision always comes at a price (Figure 6.8).

To reduce costs, it sometimes makes sense to adjust the dimensional tolerances by increasing the amount of allowable variation (in essence by reducing the requirements for precision). This reduction in precision will almost always result in lower processing costs—for the individual parts. It may reduce costs for assembly—if the parts now fit together easier and faster—or it may actually increase costs—if parts do not fit together right. Finding the level of precision with the lowest overall cost can be challenging.

6: Material Selection Based on Cost 235

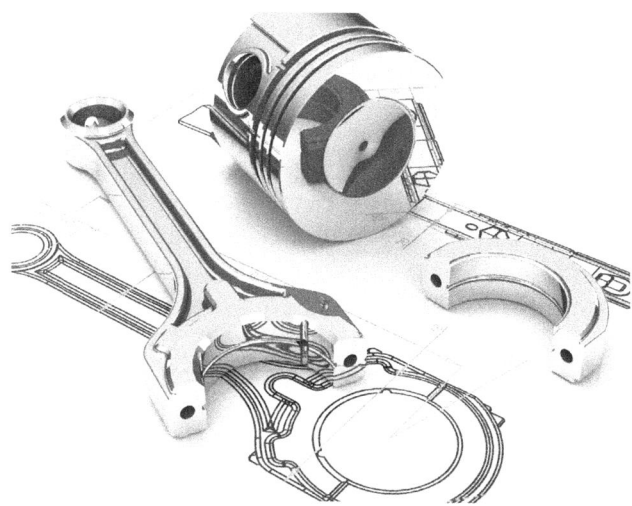

Figure 6.8 Precision always comes at a price. BildagenturZoonarGmbH/Shutterstock.com.

The concept of acceptable variation normally means acceptable dimensional tolerances. However, there may also be variations in shape and form, sometimes even in function, that are unacceptable in parts made of metal, but are acceptable in parts made of thermoplastics. Thermoplastic parts often have a unique paradigm for these kinds of variations. Among other things, thermoplastic parts are usually lighter and more flexible, which can allow for minor adjustments at assembly, allowing a low-precision part to fit just as well as a high-precision part. Thermoplastic parts are also typically softer and more ductile than most metal parts, so that even if they have a higher variation in their overall shape and size, the overall fit at final assembly is still very robust. The key question thus becomes, *What variations are acceptable?* Not just in the magnitude of the variation, but in the manner and type of variation.

As an example, let us take the classic case of a round pin that is being inserted into a round hole. Assume that our pin is precision ground from a high carbon steel, and our hole is in a precision machined part made of 17-4 stainless steel, which has been hardened. Both our pin and our hole are almost perfect cylinders, as close to being perfect as is physically possible. We may make the hole diameter a few thousandths of an inch larger than the diameter of the pin, and add a generous chamfer to help guide the pin into the hole. We might even round the end of the pin to guide it even better. We could design a pattern of such holes and pins, knowing that the flatness and stiffness of the mating parts (combined with a little bit of machine oil) will allow everything to come together in a smooth and precise manner. All of our design decisions—including material choices, production methods, and production tolerances—are predicated on our desire to have this system come together exactly as we intended it.

Now let us assume that our pin is made of thermoplastic and it is injection molded. So, we know it cannot be a pure cylinder with a perfect diameter, as there is going to be some amount of draft added somewhere—regardless of how we orient the part in the mold—and perhaps even some mold mismatch. Then we have a hole in another part, which is thermoformed, or perhaps rotationally molded. That hole is not going to be a perfectly round hole (or straight), no matter how much we may want it to be. Now, we have to establish size targets for the pin and for the hole, evaluate the tolerances for a typical production scenario, and then do a tolerance analysis to ensure that everything always fits.

If we have a pattern of pins and holes, the analysis might get even more complicated. Our design intent—which is to have this system come together in a smooth and precise manner—has not changed. However, the materials have changed, the production methods have changed, and our production tolerances have changed. The materials we are using are now softer and more flexible, and it may very well be the case that they fit together better than before, even though the dimensional variations among the individual parts are greater.

This concept of acceptable part variations—beyond the variations due to dimensional tolerances—can often allow a design team to create a lower cost design solution, not just because of lower material costs, but because of lower processing costs. Unfortunately, it is a concept that is difficult to measure and quantify. One tool that is widely used to define dimensional tolerances is a system known a geometric dimensioning and tolerancing, or GD&T. GD&T is a comprehensive methodology, and properly applied it can be used to monitor and control not just linear dimensions (i.e., size), but overall shape and form as well. However, while GD&T can be used to define acceptable part variations, we first need to understand what those acceptable variations are. To do that, we need to go back to the basics of working with thermoplastic materials.

As we have discussed throughout this book, thermoplastics are unique materials. And while you can study their behavior mechanical, and calculate their deflections under load—in either static or dynamic loading (or both)—to work with them effectively you have to become familiar with them. You need to understand how they bend, flex, twist, and warp, how they respond to being mated and/or constrained to other parts with

6: MATERIAL SELECTION BASED ON COST

different stiffnesses and different expansion characteristics. The reality is that the form and fit of a thermoplastic part, or a set of thermoplastics parts, is often very different than the form and fit of parts made of metal. For those of us who grew up working with metals (in other words, most of us), it takes time to assimilate this concept.

Our initial reaction is often to treat thermoplastic parts the same way as metal parts. We create a design, and then specify very precise dimensional tolerances—because we can. And of course we will highlight dozens of different dimensions as critical dimensions—because all of them will be critical. And we might even create a clever design solution for mating parts using locating pins on one part, with holes and slots on the other part. But even with a clever design solution, we need to remember that plastic parts often fit together in different ways than metal parts. If we understand how they fit, we can often allow for more variation—in both dimensional tolerances and acceptable shape—and thus reduce costs.

Years ago, I was working on a cell phone project for Nokia. Like most Nokia phones, this phone had removable covers, which allowed the user to swap covers and change colors as they desired. However, this phone also had a camera and the camera bezel was mounted not to the frame, but to one of the removable covers. My job was not only to design the camera bezel and the cover, but to make sure the camera bezel always lined up with the camera, no matter what.

To attach the bezel to the cover, I came up with a series of holes on the bezel and a series of pins on the cover. The bezel would slip over the pins, and then each pin would be heat staked (basically melted in place), and the staked pins would act as rivets to retain the bezel. The entire cover assembly would then get snapped in place on the back of the phone. Early in the design phase, we went through an extensive tolerance analysis to identify all of the variables that could affect the final location of the bezel, and keyed in on series of specifications for the pins, including some high precision tolerances involving true position and features of size. However, when we started experimenting with actual parts, we found the exact position of the pins was not as critical as we thought. They were small, flexible pins and they could easily adapt to any mismatch (and they would get melted at assembly anyway). All we needed to do was locate the bezel in the center of the opening (easily done using chamfers and angles) and then heat stake away. Afterward the cover would snap into place on the phone, and self align itself, even if there was a bit of warp or twist in the cover part. If these parts had been made of metal, they would have needed a completely different level of precision (Figure 6.9).

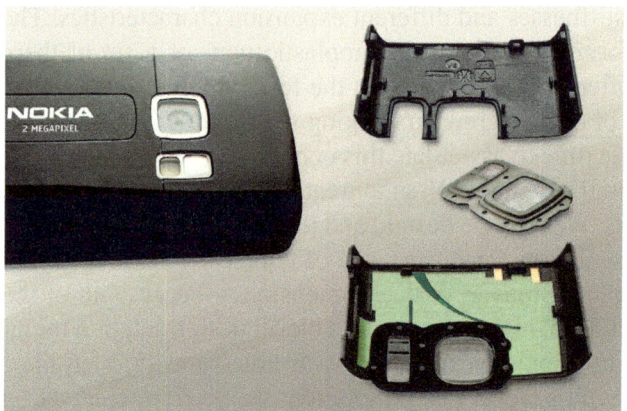

Figure 6.9 Camera bezel parts and assembly.

6.6.3 Uniform Wall Thickness

To further reduce processing costs, one should always design plastic parts to have uniform wall thicknesses. This applies to almost all plastic processing methods, but has special importance for injection molding. There are several reasons for this.

First and foremost, uniform wall thicknesses allow for uniform flow and fill of the molten material in the mold cavity. Among other things, the flow of the material affects the orientation of the polymer chains. The more uniform the flow, the more consistent the physical properties of the material throughout the part. Uniform flow also allows for uniform stress distribution throughout the part. All of this can affect the structural performance of the part, its physical dimensions, and even its appearance.

Second, uniform wall thicknesses allow for uniform cooling. Cooling time is often the longest part of the injection molding cycle. During this part of the cycle heat is transferred from the molten material in the cavity into the surrounding steel. The rate of heat transfer may vary through the part. It is dependent on the shape of the part, the materials used in the mold, and the layout of the mold and its pattern of cooling lines. The rate of heat transfer will affect the shrinkage of the material, and in the case of semicrystalline materials, can also affect the rate and overall amount of crystallization, further affecting the shrinkage. A part with uniform cooling will have uniform shrinkage, which will result in more consistent part dimensions. In essence, uniform wall thicknesses make the production of higher precision parts (tighter tolerances) a little bit easier.

So how does this reduce processing costs? By making something easier to do, you are also allowing it to be done faster. Not only will the process cycle time be less, but the time and effort to set up the process will be less.

Both of these reduce costs. In addition, since the parts will be structurally sound, dimensionally accurate, with the desired properties, the long-term product performance will be maintained, and the downstream costs (for product returns, field failures, loss of sales due to quality issues, etc.) will be lower as well.

In addition to having uniform wall thicknesses, one should also try to minimize the wall thickness whenever possible. A thinner nominal wall thickness will reduce the required cooling time in the mold, and allow for the fastest possible molding cycle for a given part geometry.

It is helpful to remember that uniform does not mean identical. The word uniform is derived from the Latin words *unus* (meaning "one") and *formis* (meaning "having the form of"). While the wall thickness does not have to be identical everywhere, the structure should have walls that are of a consistent shape and form. Other concepts to keep in mind involve symmetry, proportion, and balance.

In the images below, there are several variations of five-legged chair bases. Each base is injection molded (in a large, single cavity mold running on a large injection molding machine), from glass-reinforced nylon. The glass loading is usually in the 30–35% range, and the type of nylon is usually 6 nylon, although, 6-66 copolymer blends are sometimes used as well. Each of these chair bases has a slightly different geometry, with different leg profiles, different wall thicknesses, and different rib patterns. Interestingly enough, they all weigh about the same, and the material costs for each design is about the same. However, the molding cycle time is very different for each one. This is partly because of the flow of the molten material in the mold cavity, which is aided and/or restricted by the wall thickness and the ribbing pattern. This affects fill time and total cycle time. More importantly, the hold time in the molding cycle is dependent on the overall wall thickness. The thicker the part, the longer the hold time, the longer the overall cycle time, and the higher the processing cost. The white chair base has the lowest overall cycle time. By no coincidence, it is also a well designed part (Figure 6.10).

6.6.4 Exploit the Material

You can also reduce processing costs by exploiting some of the nuances between different materials. Just as you might select a certain paint because it dries faster, you may sometimes select a thermoplastic material simply because it processes faster. A certain material might flow better, which allows for thinner nominal walls, which then results in a shorter cycle time because it cools faster. Or a certain material may shrink in a more uniform manner, allowing for more precise parts, which reduces assembly cost.

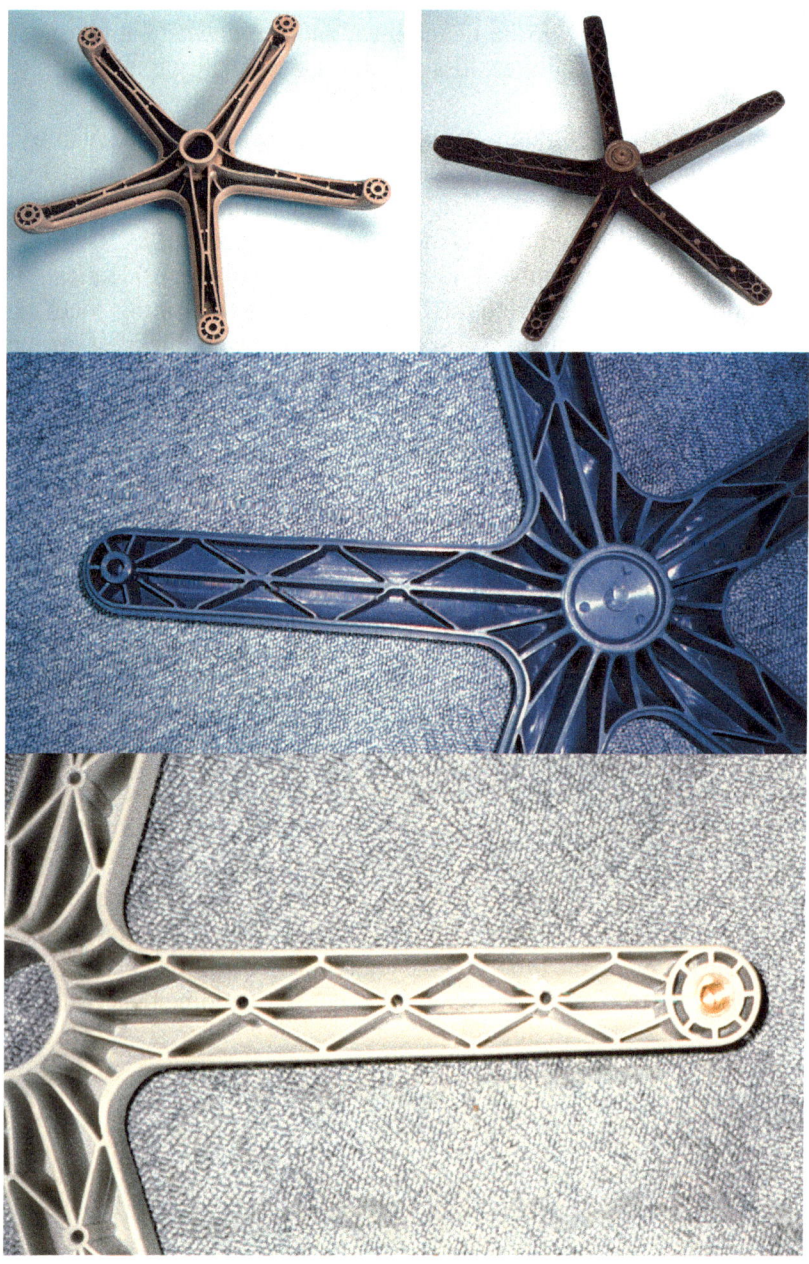

Figure 6.10 A variety of chair base designs.

Sometimes there are two materials that seem equally suitable, but you are wondering if either one has a processing advantage. If there a choice between an amorphous and semicrystalline resin, use the semicrystalline resin. Since it has a distinct melting point, it will become structurally stable at a higher temperature, and can therefore be ejected from the mold faster, so the molding cycle time will be less. (Note: the mold shrinkage will be higher.) When choosing between two similar materials, and all else being equal, choose the material with the highest melting temperature. Since it has the highest melting temperature, it also has the highest freezing temperature, and will solidify faster, allowing for faster cycle times. These nuances can also be exploited to reduce the cycle times of secondary processes, such as ultrasonic welding and heat staking.

These are subtle tricks, and may not always work. Even when they do, you may only shave fractions of a second off the cycle time. While a fraction of a second may not seem like much, the seconds mean pennies, pennies turn into dollars, and the dollars add up.

6.6.5 Design for Speed

Finally, there is *the need for speed* [9]. Not in terms of time to market, but in process cycle times. In every design decision you should consider the processes that are involved, the process rates, and the cycle times of these processes. Whenever you can, design parts and select materials to allow for faster cycle times. Design parts so that they can be easily handled and easily transported. Remember, the easier it is to do something, the faster it can be done.

Design parts so they can be easily assembled, which also means they can be quickly assembled. A snap-fit attachment is almost always faster than a design with a threaded fastener. A self-tapping screw can often be installed faster than a machine screw into a threaded insert.

For parts that go through feeders or hoppers—consider using a material with high lubricity (a low coefficient of friction). These parts will move faster and will jam less often. When choosing between two materials, and all else being equal, choose the material with the highest hardness. Those parts will be less prone to cosmetic damage during handling, so they can be handled faster and more aggressively (and your scrap rate due to cosmetic damage will be lower).

Consider whether the parts can be shipped in bulk (in a bin, or a box, or a bag), or will they need to be handled individually, or in a special orientation, or with a little bit of extra care? Parts that are shipped in bulk can be handled much faster.

This focus on speed may only save a few seconds here and there, but seconds mean pennies, pennies turn into dollars, and the dollars add up.

Finally, whenever possible, eliminate processes. You can spend all kinds of effort to make a certain process as efficient as possible. But the lowest possible cost scenario is when you eliminate the process entirely.

> There is nothing quite so useless, as doing with great efficiency, something that should not be done at all.
> **—Peter F. Drucker, Management Consultant**

6.6.6 Integrating Team Input

In this section, we have discussed some different ideas for reducing processing costs through more effective material selection. While some of these ideas are fairly basic, some involve special insight, and others may involve fundamental changes in how a company goes about developing its products.

There is no perfect training curriculum that will teach you how to design with plastics, and how to optimize your dimensional tolerances to get more cost-effective design solutions. With time and experience, you will begin to appreciate the subtle differences between working with metals and working with plastics, what tolerances are appropriate (not just achievable, but useful, functional, and cost-effective).

Many companies talk about having a team-based development approach, and they will go to great pains to ensure that the design team considers the input of the entire company, and/or the voice of the customer. These efforts often result in a better design solution, one that better satisfies the needs of the end user, sometimes at a lower overall cost.

What I find interesting (at times it is amusing, other times it is downright frustrating) is how few companies consider the input of the product design team in the production process itself, either in the selection of equipment, the design of the production tools, or even the layout of the factory floor.

Most design teams have people that are highly skilled in three-dimensional thinking. They can visualize how things work, how they go together, and how they might come apart. While everyone in the company wants them to listen to their ideas of how the final product should do such and such, they do not have the slightest interest in hearing design team's ideas about how the product should be built. Yes, there are some design teams that do not care, they just expect the product to be built so that it perfectly reflects their design intent. (Thankfully, I rarely work with this type of design team.) But most design teams have a keen interest in how the product gets built, and their insights and ideas cannot only be used to build a great product, but to also enable a sustainable, competitive advantage due to a reduction in processing costs.

6: Material Selection Based on Cost

6.7 Total Manufacturing Cost

In the previous sections, we have discussed some specific ideas for reducing material costs, and/or processing costs. However, we need to remember that material selection affects every variable in the part cost equation. While there may be ways to reduce costs in one area of the equation (often many ways), some of these efforts may actually increase costs in another area of the equation. The challenge to the design team: how do you evaluate all of these effects and come up with an optimal solution?

6.7.1 Explore Design Options

Sourcing teams work very hard to negotiate favorable pricing with suppliers in their supply chain. Molders know that their profit margins often depend on their ability to optimize the cycle times on their molding machines. Contract manufacturers rely on their ability to fabricate a number of components while streamlining the overall assembly process. While the common goal is to deliver a product with a defined performance at the lowest overall cost, each team sometimes become so focused on the cost reduction exercise right in front of them that they miss the overall objective. This is a classic example of the phrase, *Not seeing the forest for the trees*.

As odd as it may seem, you cannot hold a contract manufacturer accountable for the resin pricing contract that was negotiated by the sourcing team, and you cannot make a molder responsible for the 17 assembly steps that the work instructions call for. These items, and their associated costs, are a direct consequence of design: the design of the product, the design of the parts, the design of the entire system. Even though management of the total cost is ultimately the responsibility of the project manager, it all begins with design.

6.7.2 Do the Math

Unfortunately, not all designers are experts at designing parts and products based on cost, or on selecting materials based on cost. In fact, most engineering and design schools rarely teach anything about cost. Design efforts are usually focused on performance and cost concerns are on the periphery. The good news: cost is easy to learn. It is a game of numbers, involving some very basic equations. All that it takes to win the game is an understanding of the equations, coupled with a little bit of effort, and some attention to detail.

One way to evaluate the math behind the final cost is by using a spreadsheet. This spreadsheet can be simple or it can be complex (as complex as you want it to be). The spreadsheet can be designed to only calculate direct costs, or it can be programmed to also account for yield rates, tool amortization costs, etc. As with any estimating tool, the accuracy of the output will depend on the mathematics of the calculations, and the accuracy of the assumptions behind them (Table 6.3).

In the table above, we are using a simple spreadsheet to estimate the cost of goods for two different parts. For each part, the spreadsheet is used to calculate the material cost per part, the processing cost per part (based on a sum of all of the processing costs), and the total manufacturing cost. To calculate these costs, we are using assumptions about the input variables, including part sizes, resin prices, process rates, and cycle times. The value of each input variable was chosen at random, with the intent to demonstrate the basic methodology. No intent was made to mimic actual values for any particular production process or for any particular material.

One item that is not addressed in the spreadsheet is the production volume of the parts. In these examples, we are assuming the production volume is the same for each design variation, with similar tooling costs. This is not always the case. There are times when we want to estimate costs based on different production volumes. If the production volumes for each case are similar, say 10K versus 20K (or even 50K), we can probably assume the parts will be fabricated using the same production process with the same kind of tooling. However, while the equipment and the process rates and the cycle times may be very similar, the amortization of the tooling costs could be different (and there are many different methods of amortization), and this could affect the part costs. If the production volumes are substantially different, say 10K versus 10M, the tooling costs will probably be substantially different as well. Either the tools will be fabricated differently, or maintained differently, or there will be multiple tools. Regardless, tooling costs for a production run of 10M will always be higher than the tooling costs for a production run of 10K. To enable an accurate analysis, one should include an amortization of tooling costs into the cost equations.

Where this kind of analysis gets challenging is when you are trying to estimate costs using different manufacturing processes. As we discussed in earlier chapters, each thermoplastic process has its own unique set of constraints. These constraints include the size, shape, and detail of the fabricated parts, as well as the choice of materials. Each process also has a unique matrix of tooling costs and process rates and cycle times. To properly evaluate the cost implications of each design approach, one must pay special attention to all of the variables involved.

I would also encourage design teams to consider not just different processes, but also the concept of simplicity versus complexity. There are manufacturing

6: Material Selection Based on Cost

Table 6.3 A Table to Calculate Total Costs

	Part A Chair Base			Part B Plastic Buckle	
Material cost					
Material	PA 6, 30% GR	PP, 30% talc	ABS, 40% GF	Acetal homopolymer	Acetal copolymer
Price/lb	$1.85	$1.25	$1.65	$1.40	$1.20
Price/g	$0.004	$0.003	$0.004	$0.003	$0.003
Part mass (oz)	32	54	48	4	4
Part mass (g)	908	1532	1362	114	114
Material cost/part	$3.70	$4.22	$4.95	$0.35	$0.30
Processing costs					
Drying					
Process rate $/s	0.005	n/a	0.005	n/a	n/a
Cycle time (s)	10	n/a	10	n/a	n/a
Drying cost	$0.05	0	$0.05	$0.00	$0.00
Molding					
Process rate $/s	$0.02	$0.02	$0.02	$0.02	$0.02
Cycle time (s)	60	60	90	22	25
Molding cost	$1.20	$1.20	$1.80	$0.44	$0.50
Deburring					
Process rate $/s	$0.005	$0.005	$0.005	n/a	n/a
Cycle time (s)	10	10	10	n/a	n/a
Deburring cost	$0.05	$0.05	$0.05	$0.00	$0.00

Continued

Table 6.3 A Table to Calculate Total Costs—cont'd

	Part A Chair Base		Part B Plastic Buckle	
Install inserts				
Process rate $/s	$0.01	$0.01	n/a	n/a
Cycle time (s)	30	30	n/a	n/a
Installation cost	$0.30	$0.40	$0.00	$0.00
Painting				
Process rate $/s	n/a	$0.01	n/a	n/a
Cycle time (s)	n/a	60	n/a	n/a
Painting cost	0	$0.60	0	$0.00
Assembly				
Process rate $/s	n/a	n/a	n/a	n/a
Cycle time (s)	n/a	n/a	n/a	n/a
Assembly cost	0	0	0	0
Packaging				
Process rate $/s	$0.005	$0.005	$0.005	$0.005
Cycle time (s)	10	30	2	2
Packaging cost	$0.05	$0.15	$0.01	$0.01
Shipping				
Shipping cost	$0.50	$0.50	$0.02	$0.02
Total processing cost	$2.15	$3.55	$0.47	$0.53
Total part cost	$5.85	$6.32	$8.50	$0.82	$0.83

Process rates: $18/h = $0.005/s, $36/h = $0.01/s, $72/h = $0.02/s.

6: Material Selection Based on Cost

processes—CNC machining, for instance—which can provide incredible precision. There are other processes—injection molding, for instance—which can provide incredibly low processing costs. To achieve your design intent, selection of the manufacturing process may sometimes be more important than the selection of the material itself. It may make sense to select a high-precision process for one part and a low-cost process for another part. One can then integrate all of the high-precision features that are needed in the system into the design of the first part, and then integrate the lower precision features into the design of the second part. While the cost of the first part will be higher, this approach can often provide the lowest overall cost for the system as a whole.

6.7.3 Find an Edge

As we described earlier, cost is a game of numbers. While most games are fun to play, the cost game is a bit different. You do not play it to have fun, you play it to win; and in order to win, it helps to have an edge. This edge may involve an advantage in raw material pricing, lower labor rates, reduced cycle times, or lower costs for tools, machines, or other capital equipment. Or it may involve doing something a little bit better than everyone else.

In the cost game, there is no silver bullet, no magic answer that always works, no matter what. The key to success is in finding an edge and using it. It might mean having a very large mold on a very large molding machine (one that runs 24 h a day, 7 days a week, 51 weeks out of the year), with a shipping partner (with a truck and a train and a plane at your doorstep) that allows for global distribution at an unbeatable price. Or it may mean a series of smaller molds, spread throughout the world, that allow for local fabrication and local distribution on a just-in-time basis without the need for inventory, or warehouses, or shipping, or any of the associated costs. Or it may mean that you simply use standard off-the-shelf parts everywhere you can, and you simply assemble and build as cheaply as possible. (Although, in this case, you probably will not be needing this book.)

There is an old adage in the world of sports: *You cannot tell the players without a program.* In a similar vein, there is an old adage in the world of manufacturing: *You cannot reduce manufacturing costs without having a plan.* Here then, is a program—and a plan—for reducing costs through effective material selection (Figure 6.11).

6.8 A Final Word on Cost

Material selection based on cost is not about selecting a material with the lowest possible price. Rather, it is about selecting a material that—in combination with an effective design, proper processing, and

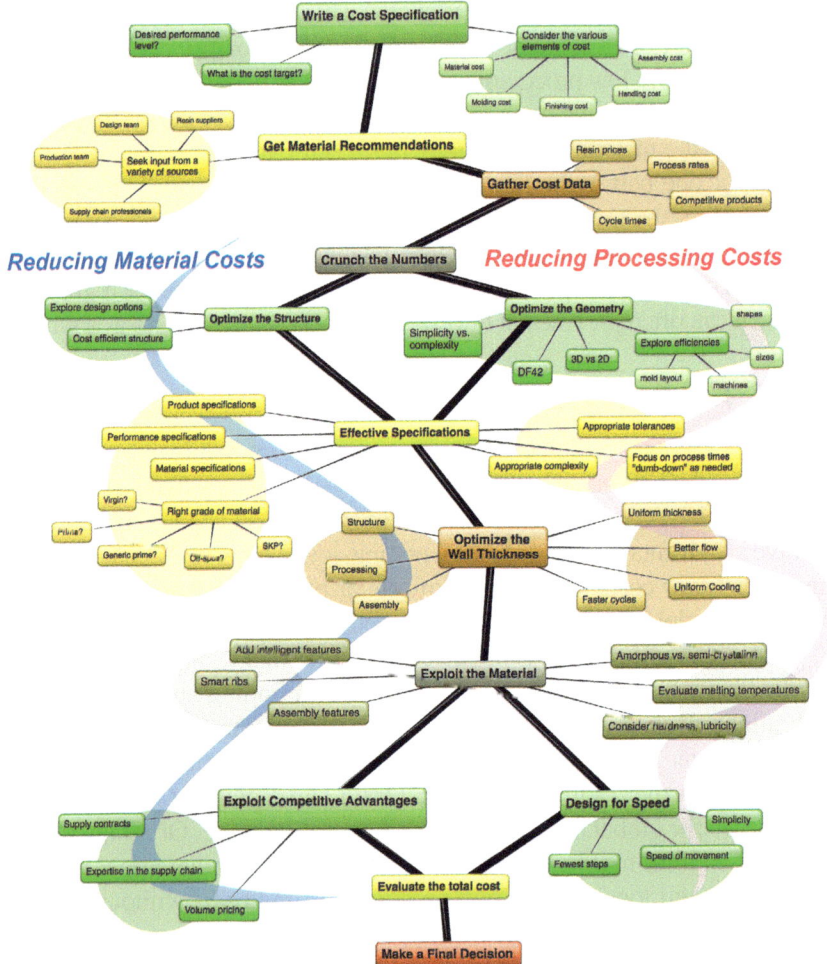

Figure 6.11 A Flow Chart to Calculate Costs. *Copyright Eric Larson - courtesy of the author.*

eventual integration into a final system—results in a product that satisfies the needs of an end user at a competitive price.

In this chapter, we have presented some ideas and tools that can guide you in this effort. How you use them, and where you use them, is up to you. Hopefully, they will help you develop a sensitivity to cost and allow you to make more effective design decisions.

If you get stuck, look for inspiration in the products shown in Figure 6.12, including a) Band-Aid® products from Johnson & Johnson, b) Velcro® hook and loop fasteners, c) plastic buckles from ITW Nexus, d) Lego® blocks, e)

6: Material Selection Based on Cost 249

the Humalog®/Humalin® insulin pen, f) plastic zippers manufactured by YKK, g) Tupperware® houseware products, and h) disposable lighters from Bic®. These are familiar items to most of us. They are well designed, cost effective, and just happen to be made out of plastic.

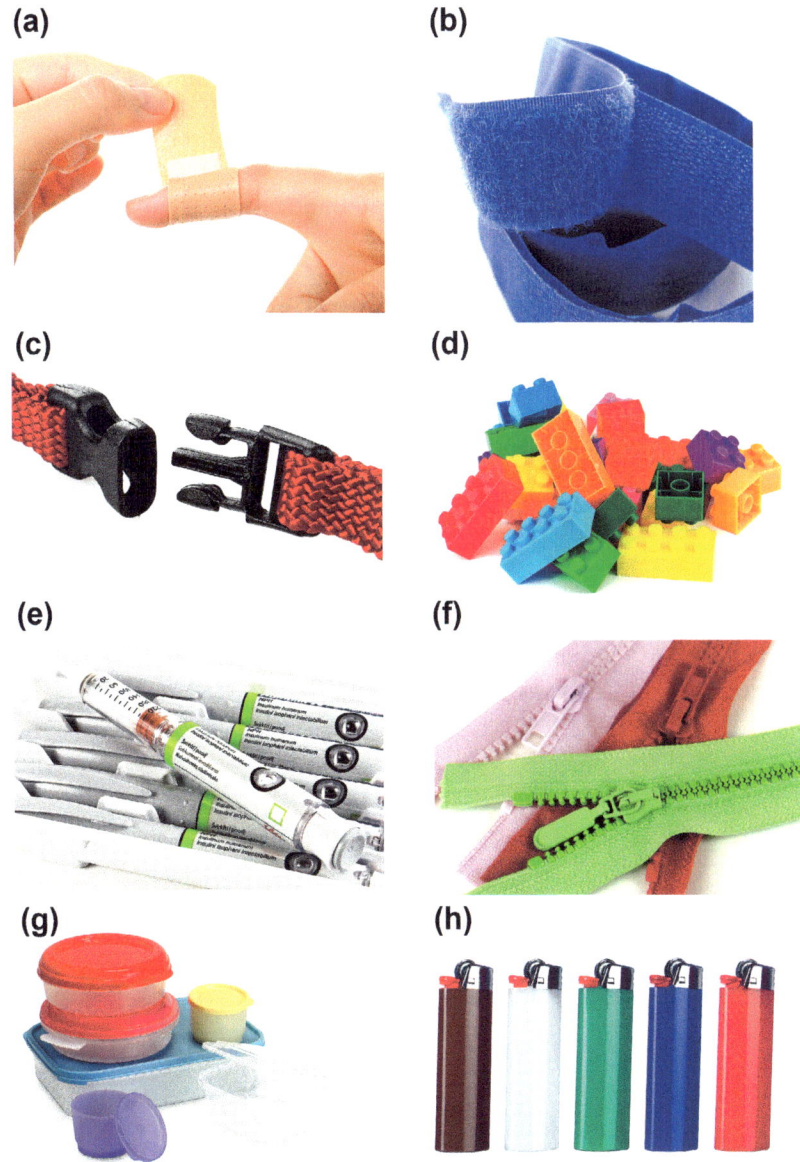

Figure 6.12 Familiar products made out of plastic.

References

[1] The International Chamber of Commerce is a Global Trade Organization, focused on policies and practices for international trade.
[2] Excerpt from the Gettysburg address by Abraham Lincoln, 16th President of the United States.
[3] The Bible, Matthew 16:26, Modern English Version.
[4] Lyrics from song "Money" from the Broadway Musical "Cabaret," produced in 1966, with music by John Kander and lyrics by Fred Ebb. In 1972 it was made into a film, directed by Bob Fosse and starring Joel Grey and Liza Minelli.
[5] Lyrics from the song "Let's Call the Whole Thing Off" from the 1937 film "Shall We Dance." Originally performed by Fred Astaire and Ginger Rodgers, words by Ira Gershwin.
[6] Loosely attributed to John Heywood, an English author, composer, and playwright. This phrase was described in his novel, "A dialogue conteinying the nomber in effect of all the proverbes in the Englishe tongue," 1546.
[7] D. Blanchard, Supply Chain Management Best Practices, second ed., Wiley, 2010.
[8] Douglas Adams, The Hitchhiker's Guide to the Galaxy, Pan Books, 1979.
[9] The phrase "the need for speed" is a slang expression that is widely used in US culture. While the exact origins are unknown, its use was popularized by the 1986 film "Top Gun," directed by Tony Scott and starring Tom Cruise and Kelly McGillis.

7 Material Selection Based on Feel

As we have discussed in earlier chapters, material selection is about finding one or more suitable materials that—in combination with an effective design, proper processing, and eventual integration into a final system—results in a product that meets its intended use, and satisfies the needs of (and hopefully delights) the end user.

One of the ways a product provides user satisfaction is through feel. This is true not only for end-use applications where people come in direct contact with the product, but also for industrial applications. This topic is often overlooked in the plastics industry, and even when there is a direct intent to select a thermoplastic material based on feel, there are few resources available. Most engineers and designers do not understand the fundamental issues involved, and very few companies have the infrastructure needed to support the process.

However, the fact remains that feel is a critical component of any product.

In this chapter we will discuss how to select thermoplastic materials based on feel. For our purposes, when we talk about the feel of a material, we are talking not only about how it feels according to our sense of touch, but also about the sensations and responses we may have when we encounter that material. In other words, this chapter is about selecting thermoplastic materials based on human response.

7.1 What Is Feel?

Feel is a simple word, yet it has a number of different meanings (Figure 7.1).

In its most common use, feel refers to the sense of touch. We often hand something to someone else and say: *Here, feel this*. Yet even in this instance, feel is more than just having something come in contact with our fingers. Our sense of touch involves a complex system, involving receptors in our skin, muscles, joints, internal organs, even our bones. This system is one of the primary ways with which we perceive the world around us, up and down, soft and hard, hot and cold, pleasure and pain.

We also use the term feel to describe what we have perceived via touch. *This feels smooth. This feels hot. This feels solid.* The descriptions are based on information we receive via various touch sensors, which are

Figure 7.1 Touch is often synonymous with feel. BardSandemose/Shutterstock.com.

Figure 7.2 Hot to the touch. DmitriMaruta/Shutterstock.com.

then correlated to our internal database of life experiences. Everyone goes through a learning process in developing that database, ranging from childhood experiences with heat (as in *Do not touch that it, it is hot*) to learning about mass and volume and density (Figure 7.2).

Feel is also used to describe sensations we are experiencing. These sensations are our response to touch, or to sensory input from our other senses: sight, smell, taste, and hearing. *I feel nauseous. I feel dizzy. I feel warm*. These kinds of responses can be visceral and powerful. Fingernails on a chalkboard. A song sung so badly out of key, it causes physical pain. Something you taste that makes you go, *Yuck*. Something you smell that makes you go, *Eww, that stinks*. And while we may call them feelings, they are not emotions (Figure 7.3).

These kinds of sensations may also involve memory and emotion. The sensory input may remind you of a past event (good or bad), or a time of the year. The warmth you feel inside when you smell fresh baked cookies. The feel of

7: Material Selection Based on Feel

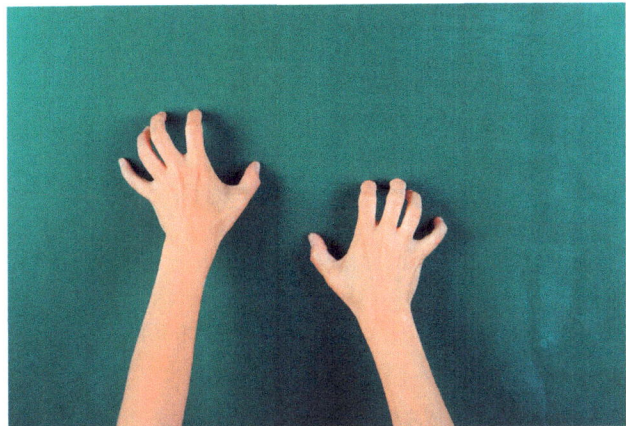

Figure 7.3 This sound often creates a visceral response. racorn/Shutterstock.com.

winter. A feeling of doom. *I have a bad feeling about this!* A feeling of being home. This response to sensory input is an important aspect of feel.

Finally, there are emotions: anger, fear, sadness, joy. And a whole spectrum of feelings in between. While a material by and of itself very rarely generates an emotion, we need to be aware that the materials used in products often affect the sensory input of the people who interact with those products. This sensory input can create physical sensations that result in an emotional response (and sometimes a very powerful emotional response).

7.2 Why Is Feel Important?

When we talk about the feel of a product, we are referring to much more than simply the texture of a surface that we touch. The feel of a product involves all of the sensory input we receive when we interact with it, plus all of the sensations we experience from that input, and all of the memories and emotions that are evoked. The feel of a product affects its performance, as well as how people interact with and use (and buy) the product. The feel has a major effect on product sales, market share, and ultimately, the business bottom line.

7.2.1 Product Performance

Feel is an important aspect of product performance. In some applications, feel is so integral to the user experience that it becomes part of the product specification. This is especially true in sporting goods and automobiles. In other applications, feel is an afterthought, something to check at the very end. If poorly addressed, this can lead to user frustration, buyer's remorse, or even a loss of sales. When questioned, users often cannot verbalize the reason for their frustration, and say something like: *It just does not feel right.* There are also some applications where feel is the only performance criteria.

One of the great challenges in material selection based on feel is in establishing the proper criteria for feel in regards to product performance. What parameters should be measured? What are the desired values for each parameter? How do each of these parameters contribute to the overall feel, and the overall product performance? We will discuss some of the material parameters involving feel later in this chapter.

7.2.2 Sales and Market Share

Feel can also provide a manufacturer with a competitive advantage in the marketplace. An excellent example of the financial power of good feel comes from the golf industry.

In the history of golf, there have been numerous examples where the equipment has been modified due to advances in materials technology. This is true not only in golf clubs, but in shoes and apparel, golf bags, even golf balls. In the early 1900s, the golf ball of choice for most players was a three-piece design, consisting of a solid rubber core, which was then wrapped with a series of rubber threads, and then covered with an outer core. This three-piece design, known as a wound ball, was manufactured for decades. The outer core was first made using a material called balata, which was a natural rubber that comes from the sap of a Balata tree. Balata balls were the standard for most good players, including professionals. They were high performance golf balls, with a combination of distance and control. The soft balata cover allowed a skilled player to put spin on the ball, usually back spin but also side spin, and they could shape the flight of the ball, control where it landed, and how it behaved on landing. However, the balata cover was not durable, and easily damaged.

In the 1960s, golf ball manufacturers also began making wound balls with the outer cover made of synthetic materials. While these materials were more durable than balata, they had a different hardness, and did not provide the same level of control to the skilled player. Manufacturers also began making golf balls without rubber windings. Some of these were solid, some were two-piece, or three-piece, or even four-pieces. They were cheaper to manufacture and much more durable. However, most skilled players avoided them. Not only did they not spin the same, they did not sound the same, they did not look the same, they did not feel the same. *It is like hitting a rock*, some said.

7: MATERIAL SELECTION BASED ON FEEL   255

Figure 7.4 Titleist Pro V1 golf ball.

In the 1980s, the Titleist Company began making a high performance wound ball known as the Titleist Professional. It had a synthetic cover as well, but was softer than other synthetic materials, and provided a performance similar to the balata ball, but with greater durability. It also sounded, looked, and felt like a balata ball. It quickly became a player favorite.

In 1999, Nike introduced a new high performance golf ball, one that did not have rubber windings. It was known as a solid core ball, and consisted of various layers of synthetic materials. Thanks to the endorsement of a young professional by the name of Tiger Woods, Nike soon began making major inroads in the golf ball market. A year later, Titleist introduced their own high performance solid core golf ball, named the Titleist Pro V1. It quickly became the most widely used ball in professional golf, and the number one selling golf ball in the world. The reasons were many: performance, durability, and feel. A Pro V1 had all the advantages of the modern solid core balls, but with the look, sound, and feel of a traditional wound golf ball. In 2002, Acushnet, the parent company of Titleist, reached $1 billion dollars in annual revenue, and the wound ball had become a relic (Figure 7.4).

7.2.3 Technical Validity

When we first began discussing the chapters to include in this book, my editor asked me a simple question. "What methods do people typically use to select thermoplastic materials?" We talked about methods to select materials based on performance, appearance, and cost. And then, without really thinking, I said something to the effect of, *You can also select thermoplastic materials based on feel.* As soon as I said that, I had a knot in my stomach. A little voice in head said, "You can't do that. Material selection based on feel is *wrong*. It's simply not...*technical*."

Over time, I began to realize that people select materials based on feel all the time. Furniture makers select wood based on its grain and sheen, and how it feels when you touch it. Products of all kinds, both consumer and industrial, have gaskets and vibration absorbers and sound dampeners. And when it comes to cams and mechanisms—the really geeky stuff—engineers often select materials based on the sound they make or the vibration they make—or do not make. All of these material choices are based on feel.

Rarely though, is there a methodology behind this selection process. I believe part of the reason is that most technical people do not want to talk about feel. If there is an analytical process involved, maybe. There is a positive bias to thinking—which is a rational, valid process—and a negative bias to feeling, which is considered an irrational, invalid process. However, I believe thinking and feeling are both valid processes, and either side of the brain (and hopefully both) can be used in the selection of thermoplastic materials.

7.2.4 The Bottom Line on Feel

Feel is the final hurdle in effective material selection. After all the performance measurements have been taken, and all the cost analyses have been done, the ultimate determinant on the success of a product often comes down to feel. This is hard for many people to accept. As Spock might say, it is not logical.

The funny thing is, despite our insistence that we are rational, logical creatures who go about our daily lives making conscious choices, most of us live by feel. We engage in activities that make us feel happy. We choose friends who make us laugh. We buy clothes that not only look nice, but also feel comfortable. We buy toys and games and cars (especially cars) that make us feel good.

Feel is real.

7.3 The Language of Feel

A major challenge in selecting materials based on feel has to do with language. Feel is a complex subject, involving multiple senses and body sensations, but also memories and emotions. Describing these issues may involve words and terms that are ambiguous, even downright confusing. The word *feel* also has many different meanings.

For most people, *feel* is synonymous with *touch*. When someone says, "Here, feel this," what they are really saying is, "Here, touch this." Touch is an important sense, and plays a critical role in feel. But even if we define feel based solely on our sense of touch we may overlook a number of important factors.

As an example, let us look at the feel of something with a light texture. Let us say you gather six samples of different materials, each one carefully selected so that it has an identical physical surface to the next—the exact same surface roughness, waviness, etc. Included in the sample set is a piece of concrete, a piece of velvet fabric, a piece of sandpaper, a machined mold surface, a molded plastic part, and an iPad. To make the exercise even more interesting, you put the samples in a climate controlled room, and let them sit for a period of time so they are all the same temperature (Figure 7.5).

Then, you lightly touch each sample with your fingertips, and move your fingertips across the surface of the sample. An interesting thing will happen: each sample will feel different. Six identical surfaces, yet six different responses. Why is this? The reason is that your fingertips are receiving sensory input that involves more than just surface roughness. There is thermal conductivity, electrical conductivity, an assessment of hardness, and more. And if you were to pick each sample up, there would be an assessment of weight and density, how vibration travels through the sample, and more. Again, each sample would feel different.

7.3.1 Sensory Input

Understanding this concept of sensory input is the first layer in the process of material selection based on feel. Although, when it comes to materials, we are not only talking about the sensory input we receive through our sense of touch, but through any of our senses. While feel is technically based on the sense of touch, the mechanism of stimulus and response is similar for all of the senses, and the process of material selection is the same, regardless of which senses are involved. Rather than describe this process as *Material Selection Based on Sensory Input*, we will use the simplified phrase *Material Selection Based on Feel*. In this chapter, we will address only the five traditional senses: touch, sight, hearing, smell, and taste.

The language of feel in this first layer is often over simplified. We may ask questions like, *Did you hear that? Can you smell that? Do you like the taste?*

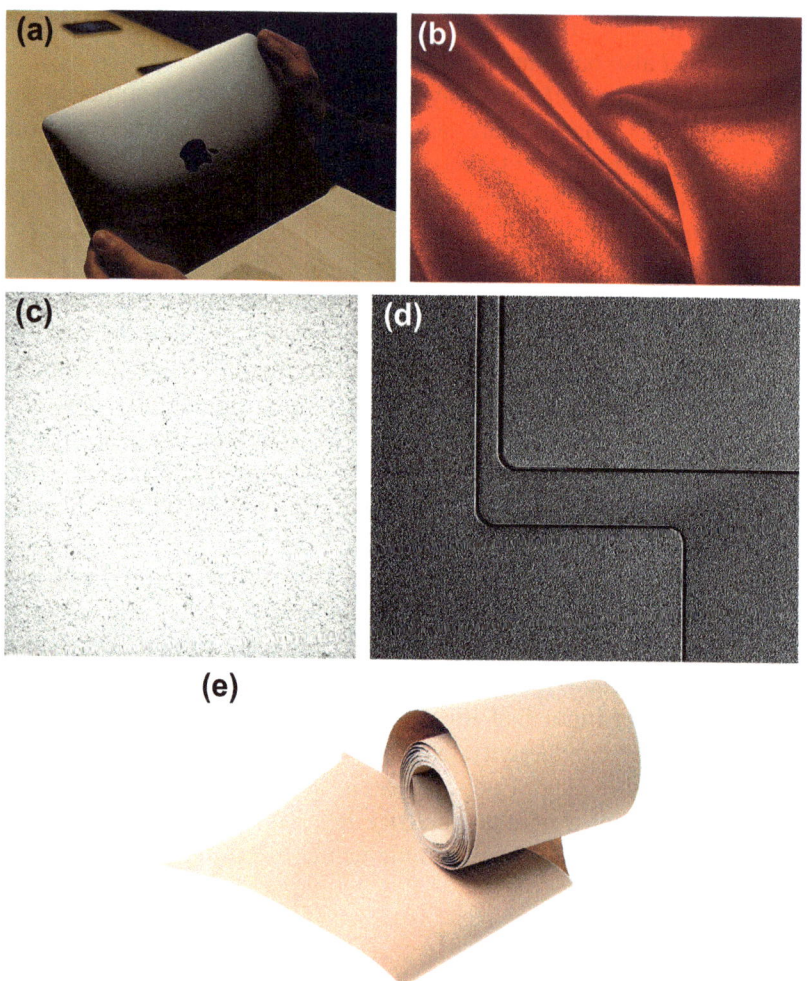

Figure 7.5 Same textures, different feel, a) iPad, b) velvet fabric, c) textured concrete, d) molded plastic, e) sandpaper. Matt Buchanan/ Wikimedia Commons, DoheeHan/Shutterstock.com, scyther5/Shutterstock.com, Prapann/Shutterstock.com, vadimmmus/Shutterstock.com, *and Photography By* MK/Shutterstock.com.

We are not asking questions like, *Have the sensory receptors in your skin detected an electrical charge in the air? (I think it is going to rain.)* This kind of language, while accurate and specific, is not something most of us in design and engineering are accustomed to. It is more common in fields like biology, physiology, and psychology. However, we can learn from these other fields, and utilize their findings and insights to make more effective choices on material selection. First we have to learn a little bit about the language they use.

7: MATERIAL SELECTION BASED ON FEEL

The Esalen Institute [1] is a retreat center in Big Sur, California which offers a variety of workshops in psychology, personal growth, and various topics at the intersection of the humanities and sciences. Some years ago, I had the experience of being a work scholar for several months. Each morning, our work crew would take a break, sit around in a circle, and have a short discussion of the day's events. As part of that discussion, we would go around the circle and each person would take a minute or two and check-in using what was called a weather report. That weather report might include a comment on how they were feeling at that moment, not just happy or sad, or warm or cold, but whether they had a cold, or if they had not been sleeping well, or if they were working on some personal issue in their life that was affecting their energy level. It was a personal experience, not necessarily intimate, but very personal. It was an opportunity to connect not only with your own sense of self, but also with each other.

In one of the work crews, there was a young woman who would begin her weather report with a description of what was going on with her physically. Not, "Oh, my back hurts" or "I am sore from yesterday," but a kind of simple status report on what her body was sensing at that moment. It sometimes included comments about the warmth she felt on her face from the sunshine, or the coolness on her arms from the wind, or the tingle she felt in her buttocks from the cold stone she was sitting on. The first time she did it, I thought it was a bit weird, but over time, I began to find it intriguing, and I started to do the same thing. It was an insightful experience, and it heightened my awareness of the sensory input was body was receiving. (At the same time, it also raised my awareness of why people often respond to the question, "How are you feeling?" with the simple answer, *I am fine*. Few of us have the patience to listen to a lengthy weather report from everyone we encounter.)

7.3.2 Human Response

The next layer in material selection based on feel involves the concept of response. Like all animals, humans respond to sensory input. The response may be an involuntary action (like your knee kicking out), a body sensation (feeling dizzy), a physical action (salivating when you smell food), an emotional response (feeling happy when you hug your child), or a combination of these items. It may also involve an intellectual assessment, where one evaluates the sensory input and makes a conscious decision on what actions to take next (such as whether or not they should yell "Turn that music down!"). These responses may also be affected by conditioning (either intentional or unintentional). The Russian physiologist Ivan Pavlov

Figure 7.6 Ivan Pavlov. Does that name ring a bell?

is well known for his pioneering work in classical conditioning, where he conditioned dogs to salivate at the ringing of a bell [2] (Figure 7.6).

Regardless of the actual response, the important thing to remember is that all sensory input results in a response, and that material selection can influence that response. We will discuss some of those influences later in this chapter. Thermoplastic materials offer unique opportunities to affect human response—not just because of their unique behavior, but also because of the capabilities to modify and enhance various properties through the use of additives. We can modify a thermoplastic material to do almost anything, and many of those modifications can also enhance a desired human response.

Unfortunately, predicting the given response to a given set of sensory stimuli is not always possible. Responses will vary depending on the individual, the type of input, even the time of day. As an example, let us suppose you set the temperature of a large room to 23 °C (73 °F)—which is considered a comfortable room temperature, according to most sources. You then invite a number of people into the room, and see how they respond. It is almost certain that there will be different responses. Some will say it is too hot, some will say it is too cold, some will complain, and not everyone will be comfortable.

Even if everyone had the exact same response, they will probably describe it in different ways. This is a key challenge for the language of feel: what words do we use to describe our response? And how do I know

7: Material Selection Based on Feel   261

Figure 7.7 An assortment of vacuum tubes.

the response you are talking about (the feel as you experience it) is the same response I am talking about (the feel as I experience it)?

The use of vacuum tubes in audio equipment provides an excellent example of the language challenge regarding our responses. With the goal of getting the sound of amplified sound "just right," many audiophiles and musicians often use vacuum tube-based amplifiers, claiming they produce a more natural and satisfying sound than solid state amplifiers. While most amplification methods have some amount of signal distortion, tube amplification (also called valve amplification) has unique distortion characteristics (Figure 7.7).

Many tube designs have intentional distortions, created to enhance or adjust the amplified sound. The resulting differences in the sound are readily apparent, especially to a practiced ear. Many audiophiles—ranging from serious hobbyists to professional recording engineers—spend countless hours and thousands of dollars scouring specialty shops and online auctions to build vacuum tube collections. Their collection might contain a mix of tube types, from different manufacturers, some selected based on specific manufacturing runs from specific dates at specific sites. There is even a subculture of companies manufacturing remakes of classic tubes, using the old specifications but with more modern processes. Tube junkies—as they are often called—will describe the sound characteristics of these tubes using words like *warm, balanced, smooth, mellow, fat,* and describe their

Figure 7.8 Wine and fruit.

dynamic range and attack characteristics. But rarely do they describe their own personal response. Instead, they focus their language on a description of the sound. Many of them do not even realize that their description is based partly on their own experiences and expectations.

A similar phenomenon occurs with responses to smell and taste. If you have ever been to a wine tasting, you have heard people describe the flavor of the wine they are tasting. Perhaps you might hear someone say, "It has a light buttery flavor, with hints of apple and cinnamon and just a touch of citrus. It has a slightly acidic finish, and the fruit dissipates into a ginger spice." If you have have not tasted the wine yourself, your first reaction is something like: *What the hell are they talking about?* However, if you have tasted the wine, and are sharing the experience with that person, you will begin to engage in a dialogue. *Oh, it is not buttery, it is a soft caramel flavor, and yes cinnamon and citrus is there, but it is not really apple, it is more like pear.* "Oh, yeah, you're right." (Figure 7.8).

In the process of this conversation, you have taken the sensory input you have received, combined it with the sensations you experienced (including the pleasure or displeasure of the flavor and the aroma), and correlated it with prior experiences and memories. You then took someone else's description of the same experience (as described in their language), interpreted that description, and then described your experience in your language (perhaps using some of their earlier language), to eventually create an agreed upon description. This is what it is like to talk about feel.

There are times when this dialogue results in a common ground, with clear language and a shared understanding. However, there are also times where the conversation results in confusion, misunderstanding, and outright exasperation.

7: Material Selection Based on Feel

The classic comedy routine by Abbott and Costello, *Who's on First* comes to mind.

7.3.3 Not an Engineering Language

For many of us in the technical world, engaging in this kind of ambiguous dialogue can be frustrating. We want accuracy, precision, and specificity. We want to be able to describe something in a way that it can be accurately measured. But how does one measure "good feel"? An equation that I derived is given below.

$$\text{Good feel} = f(\text{design, material, application, } \mathbf{WOW} \text{ factor})$$

where

$$\mathbf{WOW} = \frac{\text{overall utility}}{\text{product cost}} \times (\text{huh})^n$$

and

$$\text{huh} = \sum (\text{functional, cool, hip, funky, trendy, excellent} \cdots)$$

$$n = f(\text{what?})$$

Obviously, these equations have no definitive solution (I believe the technical term is *indeterminate*). They have variables that are ambiguous and confusing, and even if their meaning was understood they are impossible to measure.

A much more meaningful equation is given below.

$$X_{tl} < X_{jr} < X_{tm}$$

where

$$X = \text{the "feel" factor we are trying to determine}$$

and

$$X_{tl} = \text{too little X}$$
$$X_{tm} = \text{too much X}$$
$$X_{jr} = \text{just right X}$$

In this equation, rather than solving for a specific value of feel, we are comparing values, and making a determination of "greater than" or "less than." This equation is based upon the work of the British author and poet Robert Southey, the author of the classic children's story *Goldilocks and the Three Bears*.

Figure 7.9 Comparative analysis.

In this story Goldilocks is a young girl who encounters the empty home of three bears. As she explores the bears' home, young Goldilocks first tries a bowl of porridge that is too hot, then one that is too cold, and then finds one between the extremes that is "just right." The theme repeats via furniture that is too big or hard, too small or soft, and then "just right." Goldilocks is making her choices by using comparison as a means of making decisions. At its heart, it is a story about material selection based on feel (Figure 7.9).

This type of comparative analysis is something that all of are us are familiar with. Ever since we were children, we have been comparing things. It is also a methodology that is widely used in the technical world. We have Go-No Go gauges, Pass–Fail testing, Accept/Reject quality criteria—all of which are based on comparison. As a species, humans are very adept at comparison. It may very well have its origins in comprehending the difference between predator and prey. One might even say that comparison is in our DNA.

7.3.4 An Imprecise Language

So how does one discuss feel? First, it is important to understand that feel has multiple aspects, some of which are interrelated, and some of which involve factors we may not be aware of. Second, we need to understand that people may be discussing different aspects of feel and not even realize it. Some people may be discussing sensory inputs, while others are discussing responses, and others who think they are discussing one thing are actually talking about the other.

On top of all that, the language is ambiguous and confusing, the responses are variable and unpredictable (perhaps even nonrepeatable), and everything is subjective and imprecise. Once you accept all this fuzziness, discussing feel actually becomes quite easy. You take your best effort at describing something, and then discuss it, make evaluations, and modify your language as needed. This is, after all, how humans have communicated for thousands of years.

Figure 7.10 Comparing apples and oranges.

7.3.5 Comparative Analysis

The important thing to remember is that feel is all about comparison. You are merely taking a set of sensory inputs, assessing the responses, and then comparing those responses to other responses. It is a very simple thing to do—so simple, even a child can do it. (Understanding how material selection can affect those responses is an entirely different matter.) The results of those comparisons can be recorded, perhaps even assessed some kind of numerical value, but the results are subjective (Figure 7.10).

There is a common expression: You cannot compare apples and oranges. It is meant to imply that one can only compare things that are similar, and apples and oranges are so different that any comparison has no meaning. I disagree with this. You can easily compare apples and oranges using a variety of criteria, perhaps based on their size, or which one you have a taste preference for, or which one you would rather throw at your mother-in-law. Comparing dissimilar items can be done, as long as you can find useful criteria for evaluation. (There are of course, situations where you cannot find useful criteria for comparison, in which case the expression may actually apply. Although, I much prefer the expression: A woman without a man is like a fish without a bicycle [3].)

7.4 Evaluating Feel

As we stated earlier, one cannot directly measure "good feel." While there may be physical properties one can measure (such as surface roughness, or density, or thermal conductivity), these properties will not be the only variables affecting the overall feel. In much the same way that performance is a subjective evaluation based on a comparison to an established benchmark, feel is also a subjective evaluation based on comparison.

To evaluate feel, one must be prepared to do comparisons. While some of this comparison may involve measurement equipment, the primary tools are the human senses.

7.4.1 Physical Equipment

Fortunately, the basic equipment needed to measure the physical properties that contribute to feel is usually simple. This may include scales to measure weight (and calculate density), gauges to determine surface roughness, perhaps a hardness tester. If thermal and/or electrical conductivity is a contributing factor, one may also have equipment to measure and quantify this—although this may not be required. Since we are making assessments based on comparison, absolute measurements are not required, simply an assessment of greater or lesser than. (This can often be done by hand, literally.)

Sometimes, the feel can involve multiple senses, such as with sound and vibration, and the measurement equipment might become a little more complex, as one needs to account for damping and energy dissipation. Again, since we are primarily making comparative evaluations, this equipment may not be required.

7.4.2 Human Senses

The primary tools for evaluating feel are the human senses. It was Aristotle who first classified the senses into five categories: sight, hearing, touch, smell, and taste [4]. In the centuries since Aristotle, scientists have added a number of new sensory inputs beyond these five initial categories. These "new" senses include thermoception (the detection of temperature), proprioception (awareness of one's body), nociception (detection of pain), magnetoception (the detection of magnetic fields), plus detection of thirst, hunger, and more. While there is little agreement on how many senses we actually have, it is clear that we have many physiological tools that we use everyday to interpret the world we live in. For the sake of simplicity, we will stick with the five basic senses for now, but will also consider the important ways in which they overlap.

7.5 Sight

The sense of sight is one of our most developed senses. It involves both the eyes, the optic nerves, and the visual cortex of the brain (located in the occipital lobe of the cerebral cortex). Human eyes contain both cone

7: MATERIAL SELECTION BASED ON FEEL

and rod cells, good for color perception and light sensitivity, respectively. The visual cortex is the largest system in the human brain, and is better developed than in any other mammal [5]. The sensory input we receive from seeing involves color, texture, patterns, depth perception, etc., and it is all based on light.

7.5.1 Light

For humans to see, there must be light, specifically light in the visible spectrum, with wavelengths ranging from around 380–750 nm. Light is a complex subject, and the technology involved is well beyond the scope of this book. With thermoplastic materials, we are primarily concerned with how they respond to light, specifically whether they are transparent or opaque, or somewhere in between.

For transparent materials, the goal is usually to have something physically present, but transparent in the visible light spectrum. We may want to bend light, but we do not want to reflect or absorb it. We may be interested in the refractive index of the material, or the total amount of light transmission, and also what wavelengths are most affected. Selecting an appropriate thermoplastic for an optical application usually involves a careful analysis of the wavelengths involved, and the required precision. That analysis is then coupled with requirements for chemical resistance, abrasion and scratch resistance, etc. (Table 7.1).

The data in the table above provide information on the optical properties of various materials. But in terms of feel—or that is to say, the human response after seeing light that has passed through a clear thermoplastic—that must be evaluated based on comparison to other materials. Architectural applications that come to mind—windows, canopies, greenhouses, office buildings, etc. How do you feel when you see light after it has passed through the clear thermoplastic in those places? Many times these materials may also be tinted, to absorb light in some parts of the visible spectrum and transmit it in others. Or they may be coated with IR filters, to prevent transmission light in the infrared region. These treatments will affect the response as well.

A few questions, for those who wear glasses. Do you feel any difference looking through glass lenses than when looking through plastic lenses? If you were wearing polycarbonate safety glasses, and you knew in your gut those glasses would protect your eyes no matter what, would that result in a feeling of comfort and safety?

There are times when we want to see through something, but we also want a confirmation that this thing we are looking through provides a safe

Table 7.1 Optical Data

Material		Refractive Index[a]		Light Transmission[a]		Useful for
Common Name	Acronym	Wavelength	Index	Wavelength	Percent	
Air		All	1.003	0.5876	1	Sustaining life
Water	H$_2$O	0.5876	1.333	0.558	83%	Sustaining life
Polymethylpentene	PMP	All	1.460	0.5876	>93%	Speaker cones, food equipment
Polypropylene	PP	0.5876	1.490	0.5876	Translucent	Packaging, clothing
Acrylic	PMMA	0.5876	1.491	0.5876	92%	Barriers, tanks, covers
Polyethylene terephthalate	PET	632.8	1.502	0.5876	87%	Containers, fibers, bottles
Polycarbonate	PC	0.5876	1.585	0.5876	88% (98% w/coating)	Automotive, aircraft
Polystyrene	PS	0.5876	1.592	0.5876	90%	Packaging, foams, cases
Window glass		0.5876	1.5–1.9	0.5876	96% w/o coating	Windows, decoration
Titanium dioxide	TiO$_2$	0.5876	2.49–2.60	0.5876	2%	Scattering light, pigments

[a] At a thickness of 3.2mm, per ASTM D103.

and secure barrier. We want to see what's on the other side, maybe even hear it (or maybe not)—but definitely not touch, taste, or smell it. So we use windows and shields and films to not only provide a certain amount of performance, but to give us a feeling of comfort and safety.

For opaque materials, color, texture, and gloss all affect how light is reflected. The resulting appearance is what we respond to. That appearance may elicit difference responses as it changes under various lighting conditions. Most interior designers know all about this, as they must consider how certain materials communicate social status, mood, and personal proclivities. Unfortunately, most plastics engineers know very little about light, and even less about color.

7.5.2 Color

Selecting thermoplastics based on color is usually done using performance-based criteria. A certain color must be within specific color tolerances and last for a certain period of time under certain environmental conditions. It is rarely done using feel-based criteria.

To understand how we respond to color it is necessary to look outside the field of engineering. In the field of psychology, there are countless books, technical papers, and research studies. There is even a field known as *color psychology*. One of the best known books is Color and Human Response [6], by the color consultant Faber Birren.

Yet for all the books and papers and theories, I am not aware of any studies that seek to evaluate the response to color as it applies to thermoplastics. What colors—and in what materials—will help elicit the response you want? Should the white aspirin bottle have a red polypropylene cap? Or should it be a blue bottle with a white polycarbonate cap? What about the Go button on your Epipen? (Figure 7.11)

In one study at the University of British Columbia, researchers found that people who had just been exposed to the color red were better at quicker recall and attention to detail, while those exposed to blue became more adept at tasks requiring creativity and imagination [7]. In sports, those who wear red are slightly more likely to defeat those in blue, perhaps because there is a powerful association of red with power or dominance, or perhaps because officials and referees treat those who wear red with more deference.

The reasons why are up for debate, but the effects of color on human behavior are measurable. Psychological associations with color are so well known that advertisers and politicians use them to advantage on a regular basis [8] and they invest a sizable amount in employing professionals to design the catchiest logos, the most empowering wardrobes, and the most sympathetic

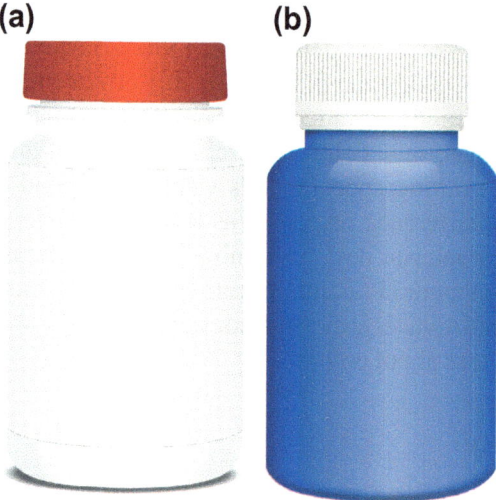

Figure 7.11 What color is your medicine bottle? a) white with a red top, b) blue with a white top

photographs. Both advertisers and artists study these effects as part of their life's work, even if they do not run clinical trials on the subject. Artists, deeply familiar with the subjective experience of shades and colors, use different language—describing a color as too warm, or too cold, for example. A color may be too saturated, and thus too intense, or not realistic enough. When artists talk this way, they are referring to the "feel" of a color, as well as the emotional responses that color—and its interaction with other colors, as well as, perhaps, the medium itself—may induce in the spectator. And actors, who often rest in the "green room" prior to a show, understand that green is a calming, creative color, an idea that has been corroborated by psychological studies [7].

7.5.3 Patterns

Another aspect of sight is the ability to discern patterns, not just due to color variations, but also due to variations in light intensity. Humans have extraordinary abilities in pattern recognition, both static and dynamic. With thermoplastics, one important kind of pattern is surface texture. The surface texture for many fabricated plastic parts is often in an "as fabricated" state, which can range from rough to smooth to highly polished, depending on the processing technologies involved. However, it is also common to intentionally create a specific surface texture, in either a random or repeating pattern.

Material selection of thermoplastics for the optimum appearance of a given texture is uncommon. While some consideration may be given the

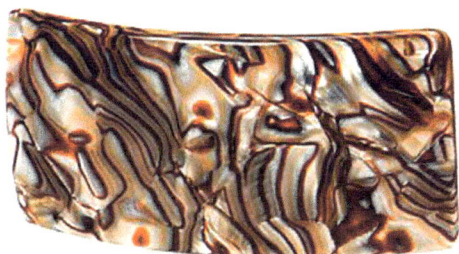

Figure 7.12 Hair barrette made from cellulose acetate.

durability of a texture, it is often due to other performance criteria. However, it is common to select materials for the durability of a smooth surface, or to add a protective coating to ensure a smooth surface. Again this is typically done as a performance issue, not as a consideration for feel.

Other patterns that apply to the thermoplastics include swirls, waviness, iridescence, pearlescence, even sparkle, and glitter. In many organic materials, such as stone and wood, swirls are often considered as a normal and natural aspect of the material. Oddly enough, with thermoplastics, swirls are often considered to be the result of a manufacturing defect. So even if that swirl is visually pleasing to the eye, the response is often an unpleasant one.

One interesting thermoplastic material where swirls are often desirable is cellulose acetate. This is an interesting polymer that is made from cellulose—plant fiber. While it is used for photographic film, a more interesting use as it relates to feel is in eyeglass frames and women's hair care products—combs, brushes, barrettes, etc. Cellulose acetate has a slight amount of transparency, as well as a certain amount of iridescence. In the hands of a skilled craftsman, parts made from cellulose acetate can mimic the look of tortoiseshell and ivory and even sea shells such as abalone (Figure 7.12).

Other interesting thermoplastics that combine colors and patterns included Corian® and Avonite®. Often called solid surface materials, these are acrylic or polyester resins that are mixed with bauxite (an aluminum ore) as well as other fillers and pigments. Material selection of these materials is based not only on the performance advantages they may offer, but also on their appearance.

7.5.4 Material Selection Based on Sight

Sight is often the sense we first turn to when encountering something new in the world around. We place a high amount of value on sensory input from sight, and rely on it to make assessments and evaluations.

Figure 7.13 Popeye the Sailor man.

And more often than not, we trust that input. Often, seeing is integral to trust. *I'll believe it when I see it*. For many, seeing is also integral to our understanding of the world around us. The phrase *I see* is often synonymous with the phrase *I understand*.

It is ironic that for many applications of opaque materials, we often try to hide the fact that thermoplastics are used. Even though seeing is synonymous with trust and understanding, we fake it. We add a brown pigment to the material and give it a wood-grain texture, or we cover it with a piece of vinyl printed with an image, or we chrome plate it to make it look like metal. Even Apple, with the iPhone 5C, which was described as being "beautifully, unapologetically plastic," went and painted every phone. It is no wonder that the word *plastic* is often synonymous with the word *fake*.

There is a movement in design and architecture that involves the concept of authentic materials. Authentic materials are materials that are presented in their natural state, that is, they are not disguised or configured to look like something else. If they are stone, they look like stone, if they are wood, they look like wood. Authentic materials are the 3D equivalent of the Popeye cartoon character, whose favorite saying was, "I yam what I yam and that's all that I yam" (Figure 7.13).

The interesting question with thermoplastic materials, what is their natural state? What should they look like to be considered authentic? The next question for the designer or the engineer or the architect: "How does the appearance of this material make a person feel?"

There are many applications where the vibrancy of an artificial color is a defining factor. The fashion industry makes extensive use of synthetic fabrics (made of thermoplastic materials like nylon and polyester and acrylic), which are then dyed. There is also nylon cord and rope, in a variety of colors and

7: MATERIAL SELECTION BASED ON FEEL

patterns. At the opposite end of the spectrum—or without a spectrum—is fishing line, where the design intent is for the line to be invisible under water. Although, I do not think anyone asked the fish how they feel about that.

7.5.5 Opportunities

What if we had a thermoplastic material that after extended exposure to sunlight, instead of changing color to yellow or faded white, it changes to a mottled grey/brown color (the color of dirt)? We would then know it is old (and dirty), but it would not stick out so much. If we could also enable a mechanism that enabled polymer decomposition at the same time, the material would fade and degrade and decompose in the same manner as organic materials like wood and leaves. Not only would it eliminate the visual blight of plastic trash, it would provide a means of returning the synthesized compound back into its basic components of carbon, hydrogen, oxygen, etc. It would also give a whole new meaning to the phrase, *ashes to ashes, dust to dust.*

7.6 Hearing

The sense of hearing (at least in humans) is based on mechanical energy. It involves the ears, with each ear having a specialized inner ear, including the eardrum. Sound waves enter the ear, reach the eardrum, and cause it to vibrate. The resulting mechanical vibrations are converted into electrical impulses, which are then transmitted to and processed by the temporal lobe of the cerebral cortex. The primary meaning of the word *audition*, which we often think of as an evaluation of one's acting or singing ability, is hearing.

While humans are fairly good at hearing, we are not nearly as adept as many other animals. The human ear can perceive a deep bass of 20 Hz all the way up to a high pitch somewhere between 20,000 and 28,000 Hz (which is not that high when compared with dolphins or bats who can hear pitches above 100,000 Hz). In addition, we can also sense mechanical vibrations through our sense of touch, including some that we cannot hear, especially when the vibrations are at the lower end of our hearing ability. In order to evaluate the effects thermoplastic materials may have upon hearing, we need to know a little bit more about sound, vibration, acoustics, and a field known as psychoacoustics.

7.6.1 Sound

Sound is a term used to describe a specific type of mechanical wave. Sound waves are mechanical waves with a frequency between

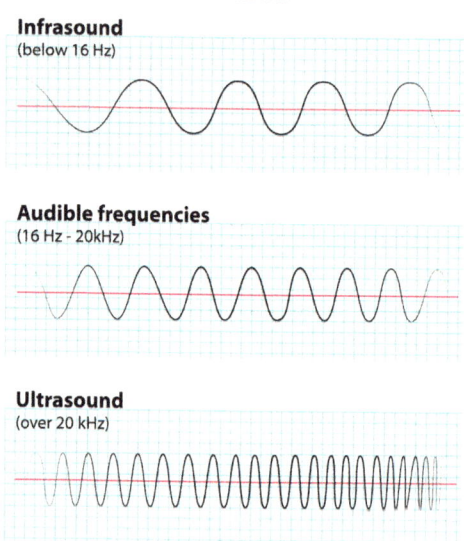

Figure 7.14 Sound waves.

approximately 20 and 28,000 Hz. This range corresponds to the range of human hearing. (In many ways, this is a circular definition, as hearing involves the reception of sound waves, and sound waves are mechanical waves that can be perceived by hearing.) (Figure 7.14)

7.6.2 Vibration

Vibration is a term that is used to describe the mechanical movement of a physical object. In a strict technical sense, vibration involves oscillations, which are predictable, repetitive movements of an object about a point of equilibrium.

Sound and vibration are interrelated. Vibration is an actual physical movement. Sound is a mechanical wave that is created as the result of a vibration. If a vibration creates a mechanical wave in a frequency that can be detected by the human ear, we will hear it, and we will call it a sound wave. A sound wave can also create vibration in other physical objects: it can cause our eardrum to vibrate, or a heavy bass beat can rattle your windows.

7.6.3 Acoustics

Acoustics is the study of mechanical waves including sound waves (audible frequencies in the range mentioned above), infrasound waves (frequencies below the audible range) and ultrasound waves (frequencies about the audible range). The field involves not only the science behind the generation of waves, but the transmission of these waves through various materials, including gases, liquids, and solids.

While most of us think of acoustics as the study of sound reflections (such as in a room or an auditorium) it is a very broad field that intersects and affects a number of other disciplines, including audio, architecture, aeronautics, even deep sea diving.

At its most fundamental level, acoustics deal with the physical production of sound waves, how they are modified and transmitted by various materials, and the effects these waves may have as they travel. Acoustical studies often include analysis of velocity, force, pressure, and other physical laws governing mechanical waves.

On the other hand, hearing also involves the reception of sound waves. The human ear—or the ear of any organism capable of hearing—is specially designed to receive these waves and then direct them to the eardrum, where they are then encoded into electrical impulses. Just as sound files are converted into formats that are more legible and efficient for use in software, neurons translate sound waves that strike the eardrum—along with other mechanical vibrations happening within the complex parts of the inner ear—into electrical impulses in a form that the brain can then receive and process. The act of hearing involves not just the reception of sound waves, but the brain's interpretation of the resulting electrical impulses. In other words, what most people think of as sound (as in *Did you hear that?*), also involves the human *perception* and *interpretation* of sound.

7.6.4 Psychoacoustics

Psychoacoustics is a field of study that explores how humans process and respond to sound, including psychological, physiological, and neurological responses [9]. It is an interdisciplinary field involving physics, psychology, neurology, music theory, music therapy, and more. The name is derived from the words *psychological+acoustics* (not from the phrase *acoustic* experts who are *psycho*). The field of psychoacoustics has numerous subspecialties, including the following:

Physiology—the anatomy of the human body, with emphasis on the human ear. Looks at the mechanics by which the ear translates acoustics into electric impulses. Determines the accuracies and limitations of the human ear.

Psychology—the study of cognition, perception, and sensation in response to sound, and what that has to do with how humans interpret certain sounds. Focuses on universal human perceptions and interpretations.

Semiotics—the linguistic, graphic, and mathematical expressions of sound.

Sociology—the social and cultural importance of sound.

History—the historical context of a sound and how that sound relates to human emotional and psychological responses.

Psychoacoustics has applications in many industries. It is especially important for designers and engineers who work with sound. While one may think they are actively involved in acoustics (the science of sound waves), more often than not they are actually involved in psychoacoustics (the human response to sound) [10]. Among other examples, engineers and software developers involved in sound compression are dealing with psychoacoustics, including the range of sound volume perceptible to humans; the sensitivity of the human ear; and sound masking, which is when a sound is rendered less perceptible due to the presence of other sounds. (Think of what is it's like when you are driving and you become aware that there is a siren nearby. While you can hear the siren, when it is coupled with the sounds of the cars around you, maybe even your own car stereo playing, it can be difficult to ascertain where the siren is coming from. This interference from other sounds is known as sound masking.) And outside of the more obvious applications, psychoacoustics plays a large role in computer engineering as well. The mechanics of the human ear engages in a translation and compression process that has been useful for developing computer networks, which rely on a similar process of packet-switching—the transmission of electronic data in packets (Figure 7.15).

7.6.5 Music

In some ways, music is a unique subset of psychoacoustics. It involves the creation of music (singing, playing instruments, etc.), the presentation of music, the recording of music, the reproduction of music, and the response to music. Musicians, artists, and producers all work with psychoacoustics on a daily basis, without even realizing it. They may be manipulating or masking a sound in order to enhance certain qualities of a recording while diminishing static, unwanted frequencies of instruments, and environmental interferences. Engineers of recording equipment and

7: Material Selection Based on Feel

Figure 7.15 Sheet music.

microphones must choose materials that dampen, absorb, project, or reflect sound, depending on the equipment's function.

7.6.6 Human Response

Music is the perfect example of human response to sound. Sound, as we discussed earlier, is a mechanical wave. Hearing is the act of receiving that mechanical wave, and how it is transmitted to and processed by the brain. The feel of a sound is the response we have to what we hear. Most of us have favorite pieces of music, which elicit physical and emotional responses, and perhaps even memories. But we also respond to other sounds, often without even realizing it.

Thump. Wack. Ka-chunck. Ker-pow. A whine. A squeak. A whistle. A grinding sound. The screech of fingernails on a chalkboard. The crumple of a paper bag. The raking of dried leaves. A hammer hitting a nail, and the ringing sound that results (and changes as the nail is driven home). The sound of someone vomiting. For each of these sounds, the individual hearing it often has a visceral, gut level response.

These responses are what we are referring to when we talk about the feel of a sound.

7.6.7 Material Selection Based on Hearing

For the plastics engineer, this brings us to the question, "How do I evaluate thermoplastic materials based on the feel (i.e., the human response) of a given sound?"

The initial answer: it depends on what your intent is. Just as we may select materials to either transmit, reflect, or absorb light, we may select materials to either transmit, reflect, or absorb sound waves.

Musical instruments are an excellent example. They are made from a wide variety of materials, including wood, metal (especially brass), animal skins, animal hair, plants (e.g., reeds), ceramics, and plastics. Each instrument has a unique kind of sound, a sonic signature if you will, and the materials that are used are carefully chosen to enhance that signature. No one makes tubas out of wood, nor do they make drums out of brass. The materials are chosen not just for their structural properties, but also for their acoustic properties—that is how the materials transmit, reflect, and absorb sound waves. The acoustic properties of a given material depend on its stiffness, density, and loss coefficient (also known as damping factor). While data on stiffness and density is readily available for almost every material, it is rare for resin suppliers to provide loss-coefficient data for thermoplastics. In addition, the stiffness of any thermoplastic changes with temperature, much more so than with other materials. (This may help explain why thermoplastics have not been a material of choice for musical instrument designers.)

In other applications, thermoplastics—especially thermoplastic elastomers—are often used where some level of sound absorption is desired. Or they may be used to alter the sound of an existing system with parts made out of metal. Since the density of thermoplastics is so much lower, the parts themselves are often lighter, so any vibration in the system will occur at a higher frequency. This may be desirable, or annoying, depending on the application. The sound of a snap—or a click, or a pop, or a thunk—is also a function of the density, and the stiffness as well. The use of additives such a glass beads, glass fibers, minerals, talc, etc., can make a substantial difference in the sound of the final product.

In order to select the "right" thermoplastic based on sound, one must make comparisons. But how many comparisons must be made? If we use our earlier example of vacuum audio tubes as an example, let us explore what tubes, or combination of tubes, is going to sound "just right" in our home music system. Let us assume that our amplifier uses three tubes in sequence, and our collection contains 100 different tubes. Assuming they are the same tube type, there are almost 1 million possible tube combinations. Are we going to test every possible combination? Of course not. But we are going to have to start somewhere. So we get a recommendation from someone—another audiophile, a dealer, an online review, perhaps a famous musician—and start there. Are we going to then do a design of experiments, or a Pareto analysis, or a linear regression? No. We are going to select a couple of different combinations, do some listening, and make some evaluations (Figure 7.16).

Based on what we hear, and how we respond to the sound (how we feel), we might swap some tubes out, or try an entirely different combination.

7: Material Selection Based on Feel

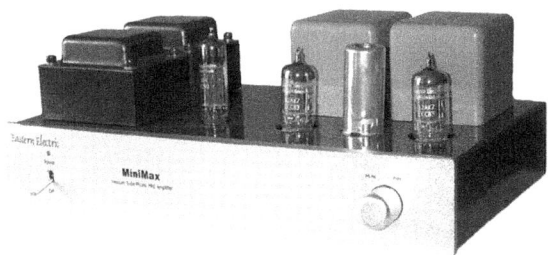

Figure 7.16 Phono preamplifier with tubes.

Pretty quickly, we will find a combination that sounds good. If you are exceptionally particular about your audio system—as I am—you may spend a few days exploring things further, but at a certain point, you are going to say, "This sounds just right." You then put the other 97 tubes back in their boxes, sit back, relax, and you enjoy the music.

Of course the tubes themselves are not the only things that affect the sound of the system. There are dozens of other variables: system set up, room acoustics, the quality of the recording you are listening to, etc. But the tubes you choose can have an effect on the feel of the sound. Similarly, the materials used in your design will affect the sound—and the feel—of the final product.

7.6.8 Opportunities

You have probably seen a plastic wine cork. Thermoplastics are being used in food packaging in a number of areas, and plastic wine corks are a relatively new application. They solve a number of issues, especially spoilage issues. And while they work technically, no one ever takes home the plastic cork as a memento of a special event. Part of that could be due to the human response of the sound a plastic cork makes when it is extracted from the bottle. What if we made the cork from a thermoplastic material that—instead of sliding out of wine bottle like a wet turd—came out of the wine bottle with a resounding pop?

7.7 Touch

The sense of touch involves a variety of different receptors of many different types located throughout the body. There are mechanoreceptors, chemoreceptors, thermoreceptors, pain receptors, and others. These receptors work together to allow us to detect not only physical shapes, but vibrations, wind currents, changes in air pressure, temperature, even electromagnetic waves.

The sense of touch also involves a complex network of sensory nerves, many of which are routed through the spinal cord into the brain, where the sensory impulses are processed by the parietal lobe of the cerebral cortex.

The sense of touch, also known as taction, is the most complex of the human senses. It is the first sense we acquire, and our first language [11]. The skin, muscles, internal organs, bones, and even teeth are all involved in touch. It is a primary method for evaluating the world around us. It is also part of how we receive and process both pleasure and pain. It is no accident that "touch" is often used interchangeably with the word "feel," and is used to describe not only the physical act of touching, but also certain body sensations and even emotional states, such as "feeling it in my bones" or when we find a sentiment particularly "touching." We often use sensory input from touch in combination with sensory input from our other senses, such as feeling sound vibrations, or examining the tactile surface of a texture while also looking at it.

Earlier we discussed how the sense of sight was related to trust. This is even more true for the sense of touch. Ironically, our sense of sight can be easily tricked, by manipulating lighting, shading, or perspective. Many of us have seen optical illusions, where the visual perception of something is different than physical reality. In a similar manner, there are also tactile illusions, where our perception of touch is different than what is actually happening. However, for the most part, we place a great deal of confidence in our sense of touch. Often times, as a final check on physical reality, someone might say, *Let me see that*. Although, what they are really saying is, *Let me touch that*.

7.7.1 Size and Shape

We use our sense of touch to evaluate the size and shape of physical objects. Often this is done in conjunction with our sense of sight, but also without. We can reach into our pocket and easily determine the difference between various coins, or we can reach for the bed stand in the dark and easily find the snooze button on the alarm clock.

7.7.2 Weight and Density

We also use our sense of touch to evaluate weight, more specifically density. We pick things up all the time, knowing instinctively how much effort it will take to do so. There is a game that is often played on young children, that begins by asking the question: *What weighs more, a pound of feathers, or a pound of rocks?* It is a trick question. It is also a revealing anecdote about the human experience (Figure 7.17).

7: Material Selection Based on Feel

Figure 7.17 Feathers and rocks.

As children, we learn that certain things, like feathers, are easy to pick up, and other objects, like rocks, are not. This learning is tempered by touching and feeling different objects—feathers, fabrics, wood, stone, steel—and our experience with the size of these objects. We may describe something as being light or heavy, but we are really speaking about the objects weight-to-size ratio. Archimedes named this "density." Most thermoplastic materials have a density slightly higher than water, but on occasion, the density can be much different.

I was at plastics trade show a few years ago, and being a plastics guy, I visited the booths of as many materials suppliers as I could. One of the distributers had an interesting presentation. One of their reps said to me, "Check this out," and handed me a small, oval-shaped part. It was about the size of a business card, and as thick as an iPhone. It had a logo, and could have passed for a decorative label. Like most plastic parts, it was warm to the touch, almost organic in its feel.

But when he let go of the part my hand almost fell to the floor. It was like someone had switched on an extra gravity field. This thing was not like a rock or a piece of steel or even a piece of lead. It was out of this world heavy. My visceral reaction was, *This thing is alien.* I can only surmise the look on my face.

"What is this?" I asked. They all laughed. "It is a new material from Ecomass. It is a thermoplastic resin with a tungsten-based filler. It has a higher density than lead, but is nontoxic and environmentally friendly."

Figure 7.18 Gel pen feel.

All I could think was, *Wow, This is really cool*. Then I started thinking about feathers and rocks.

It reminded me how every part in a system affects the feel. Whether it is the gel filled section on the barrel of a pen, or the elastomer keypad on a laptop, or the insert on the face of a putter. The density of the material in each part affects the system, the overall weight, the overall balance, the vibrational characteristics of the system, the feel (Figure 7.18).

7.7.3 Temperature

Our sense of touch is also capable of detecting temperature. This is done primarily through the skin, although the lining of our mouth (and the nerves in our teeth) can also detect temperature. We often have physiological responses to temperature (sweating, shivering, or getting goose bumps), as well as sensations of pleasure or pain, and emotional responses.

Quite often we think of temperature as it pertains to air temperature. However, we are also able to detect temperature in physical objects. Sensing temperature involves complex reactions in cells known as thermoceptors, which then send electrical impulses to the brain. Temperatures outside a certain range can also trigger pain receptors.

We often describe physical objects as being warm, cool, cold, hot, etc. What is interesting is that we may describe different objects that are at the same temperature as feeling different. This is due to the amount of heat, the thermal energy, that is transferred between those objects and our skin. The amount of heat transferred is dependent on the total amount of contact (the surface area) and the thermal conductivity coefficient (TCC) of the object. An object that has high TCC, such as most metals, will transfer heat much faster than an object with a low TCC, such as wood or thermoplastic. Below is a table with TCC values for a number of common materials (Table 7.2).

Table 7.2 Thermal Coefficients

Thermal Conductivity Coefficients of Plastics and Other Materials	
Material	W/(m × K)
Air	0.024
Styrofoam (expanded polystyrene)	0.033
Wool, blanket	0.04
Wool, felt	0.07
Wood—white pine (across the grain)	0.12
Leather, dry	0.14
Epoxy	0.17
PVC	0.19
Acrylic	0.20
Polymethlypentene	0.20
Human skin, epidermis	0.21
Acetal	0.23
Polycarbonate	0.24
Nylon 6	0.25
PTFE	0.25
Polyethylene, LD	0.33
Polyethylene, HD	0.50
Water	0.58
Human tissue, internal organs	0.45–0.58
Window glass	0.96
Concrete (dense)	1.0–1.8
Corian (ceramic filled)	1.06
Porcelain	1.50
Ice	1.6–2.2
Marble	2.08–2.94
Nylon 6, thermally conductive	0.66–4.0

Continued

Table 7.2 Thermal Coefficients—cont'd

Thermal Conductivity Coefficients of Plastics and Other Materials	
Material	W/(m × K)
Polycarbonate, thermally conductive	0.75–5.0
Stainless steel	16
Carbon steel	43
Zinc	112
Aluminum	205
Gold	310
Copper	395
Silver	425
Diamond, natural	900–2320

When we touch various objects at room temperature, we can immediately sense where they fit on the TCC scale. We know, on a visceral level, that objects made from metal will feel cooler than objects made from other materials. If we pick up a piece of silverware, or a metal case, we even anticipate that it will feel cool to the touch. There is something wired in our psyche about metals. We know they are heavier (denser), that they are cool to the touch, and we also expect them to be solid and robust. It might go back to the early humans in the Iron Age, sitting around a furnace, grunting, *metal good, make weapon*. So when we touch something made of metal, we often have a number of predictable responses.

On the other hand, most woods have a relatively low TCC. When we touch something made of wood, it feels warmer, and we respond differently. (Maybe that is where the good luck phrase, *Touch wood* comes from.) Concrete, porcelain, marble, and other stone materials have TCC values roughly midway between wood and metal, on a logarithmic scale. TCC values for glass are in the same range. Thermoplastics have TCC values ranging from a little higher than wood to about one half the TCC of concrete. Interestingly enough, these TCC values are similar to that of human tissue. While there are additives that can be added to thermoplastics to increase their thermal conductivity, they will never conduct heat the way most metals do.

A related topic that is important to plastics engineers is touch temperature. These are temperature limits, both hot and cold, for skin contact with an

7: Material Selection Based on Feel

Figure 7.19 What is the touch temperature limit for a metal horseshoe?

object in order to prevent pain and skin damage. The touch temperature limits depend on the TCC, the initial temperature of the object, and its heat capacity. Heat capacity, sometimes referred to as thermal mass, is a measurement of how an object stores thermal energy. Thermal energy involves not just the temperature of an object, but also its size (specifically, the overall mass).

As an example, if you had a horseshoe made of carbon steel, heated it red hot, and then stuck it in a bucket of water, what would happen? While some water would boil and evaporate, most of the thermal energy in the horseshoe would be transferred to the water in the bucket. The horseshoe would cool substantially, and the water would become slightly warmer. However, if you were to touch this red hot horseshoe with your finger, you would experience severe pain, and most of the thermal energy in the horseshoe would remain in the horseshoe (Figure 7.19).

Obviously this is an extreme example, but let us take it one step further. Suppose we take that same metal horseshoe, and heat it to 200 °F. Then we take another horseshoe, the same exact size and shape, but made of nylon 6, and we heat that horseshoe to 200 °F. Then you pick up one horseshoe in one hand, and the other horseshoe in the other. Which one will you drop first? Most likely, it will be the metal horseshoe. Not only does carbon steel have a significantly higher TCC than nylon 6 (over 100 times higher), the carbon steel-horseshoe has a significantly greater amount of thermal energy, due to its higher mass. That heat is going to be conducted into the skin of your hand very quickly, and you are going to drop it like a hot potato.

When we are designing objects that are going to come in contact with skin, we need to be aware of not only the TCC of the material, but the thermal energy of the system. Thermoplastic materials are used for electronic housings in a wide range of applications, including power tools and test equipment, hand-held electronic devices (cell phones, tablets, etc.), and also medical equipment. Quite often the electronics within these devices generate heat. How much heat? Where does it go? How much of this heat is conducted through the housing? What area(s) of the housing come in contact with human skin? What is the Touch Temperature Limit for the application?

Calculating the Touch Temperature Limits involves some complex mathematical equations, but as a general rule, when compared to other structural materials, thermoplastics have a lower Low Temperature Limit, and a higher High Temperature Limit.

7.7.4 Pressure

Pressure is the term used to describe the application of force over a given area. Force, which we remember from our studies on Newton's laws of motion, is an interaction which causes a body at rest to move. Our sense of touch provides us sensory input based on a little of both. We are able to detect pressure and/or force throughout most of our body. The act of detection may not have any direct correlation with the material that is providing the force. However, the detection of force (and/or pressure) plays an important role in our interaction with the world around us.

7.7.5 Vibration

As we described earlier, vibration is the oscillation of an object about a point of equilibrium. (In other words, it is an object that is moving, but not going anywhere.) Vibration and sound are related. Vibrations can create sound waves, and vice versa. Vibration can also be transmitted from one object to another. The oscillations in the first object initiate oscillations in the second, which initiate oscillations in the third, etc. (Figure 7.20).

Thump. Twang. Boing. Dweep. Clink. Boom-boom-boom. Drip-drip-drip. While these are often descriptions of sounds we hear, they are also descriptions of vibrations we feel.

Humans are able to detect vibrations over a wide range of frequencies, although the exact range depends on the individual. Most of can detect frequencies in the range of 20–20,000 Hz using our sense of hearing. Vibrations with a lower frequency are commonly referred to as infrasound, which includes frequencies beginning at 20 Hz (the lower range of human hearing) down to around 0.001 Hz. These frequencies are detected not via hearing but via our sense of touch in various parts of the body.

While it seems like we are not as sensitive to vibrations as most animals, this may be because we do not pay attention to them anymore. Also, we create so many vibrations with our transportation systems, communication lines, and media, we may actually have made ourselves less sensitive to the

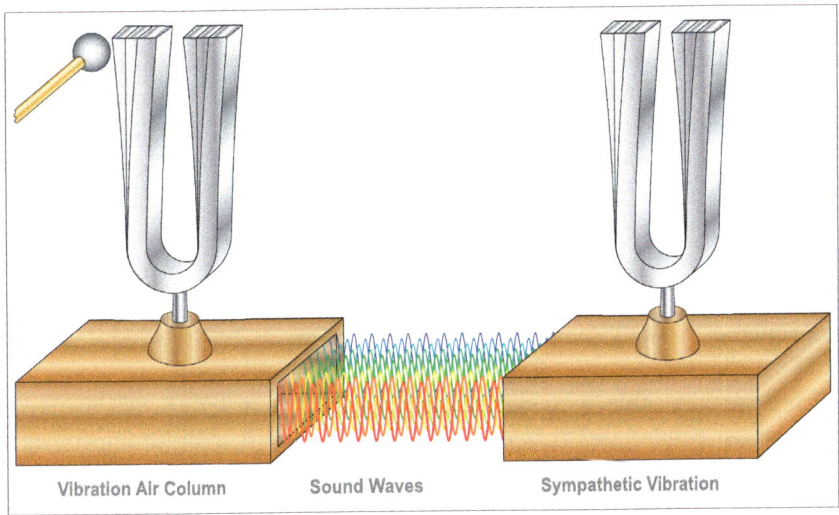

Figure 7.20 Vibration and sound transmitted by a tuning fork.

lowest frequencies. Research has shown that many nonhuman animals are sensitive to a wide variety of vibrations and may even use them for communication. For example, elephants have special sensors in their trunks that can detect seismic waves, allowing them to potentially communicate with each other over distances as far as 20 miles. Rodents sometimes drum their feet on the ground as a territorial signal. And the old wisdom that domestic animals can detect earthquakes before they happen due to their sensitivity to seismic waves may actually be true [12].

Regardless of the frequency, there are desirable vibrations (aka *Good Vibrations*), and there are undesirable vibrations. There are vibrations which make our teeth chatter, vibrations that make us nauseous, and vibrations that make our heart go pitter-patter (which is usually a good thing, but not always). There are vibrations which are resonant, consonant, dissonant, damped, undamped, muted, baffled, treated, untreated, amplified, modified, distorted, and/or ignored.

When it comes to material selection, the important thing to remember is that vibrations are mechanical waves, and all materials have unique characteristics in regards to how they reflect, transmit, and absorb the energy in these waves.

7.7.6 Movement

In addition to vibration, we can also detect other types of movement through our sense of touch. On a macro scale, we can usually detect when our bodies are in motion, whether in an elevator, or in a car or bus or train, or even an airplane. Sometimes our detection of movement is affected by sounds we hear or by the vibrations we feel, but even if we did not hear anything and could not feel any vibration we would still often know that we are moving. We might not know how fast we were moving, or what direction, but we know we are moving. Some of this comes from the vestibular system in the inner ear, and I suspect some of it comes from our ability to sense gravity, perhaps even changes in the earth's magnetic fields. We respond to this detection of movement in different ways, depending on the movements. Some movements might rock us gently to sleep, while others make us nauseous.

We are also able to detect movement in the structure of the world around us. Sometimes this is due to vibration—such as an earthquake, or a large tree falling. We feel it, throughout our entire body. But we can also detect the movement of things around us as they deform and twist under load. We've all been on a bridge—either on foot or in a vehicle—where we could sense throughout our body that the structure was different. We knew we were no longer on solid ground, we were on a bridge. Or we walked on thin ice, or squishy ground, and could feel things flexing beneath us. Or we climbed a tree, and felt the trunk and branches deform under our weight, and sway in the wind. In a similar manner we use our sense of touch to detect movement in man-made objects as we interact with them.

I recently had the opportunity to fly in a 1946 Piper J-3 Cub airplane. It is a 2-seater airplane, with a tiny little engine and wood propeller. You start the engine by grabbing the propeller and spinning it. It sounds like a go-cart. The frame of the airplane consists of a series of aluminum tubes and beams bolted together, some guy wires, and everything is covered with dope-treated fabric. The whole thing weighs about 800 lbs (360 kg) (Figure 7.21).

It was a short flight, where we took off, climbed for about a minute, then banked left, turned around, and came back and landed. They called it an *up and around*. At the highest point we were about 800 ft off the ground. We were never going that fast, about 60 mph tops. The wind was light, but could you feel the plane bounce and sway due to the wind. I could also sense the twists and flexing in the frame of the airplane itself. I could feel it through my hands, which were clenched on the cross braces in front of me. And I could feel it through the seat back and seat bottom, and on the floor under my feet. The whole thing shook and shimmied like crazy (and I had numerous visceral body reactions as a result). I later told the pilot I could feel the plane move every time he took a breath. Talk about feel.

7: Material Selection Based on Feel

Figure 7.21 Piper J-3 Cub.

The amount of movement in a structure is not always that dramatic, and while we are not always fully conscious of it, we are able to sense it. We may tell our friends about the way our car hugs the road, or the nimbleness of our mountain bike, or we may sleep peacefully to the movement of our creaky old house. We also sense the speed of response when we snap something together, and we feel the click, the pop, the catch, the confirmation that all is good. We are safe. We are secure. The door is locked. And while we may be hearing something, and perhaps sensing some sound vibrations, we are also responding to the movement of the structure. In essence it is sensory input on the rigidity of the structure, delivered through our sense of touch. The rigidity of the structure depends on its design (its shape), but it also depends on the material. We then make an assessment of whether or not the structure is too flexible, too stiff, or just right.

We also use our sense of touch to assess how things fit together. Are they loose? Are they tight? Are they sloppy? Or is everything just right? Fit often depends on the size of the things coming together, and their dimensions (including variations and production tolerances for interchangeable, mass produced parts), but it can also depend on the materials that are being used. One of the interesting things about designing for the proper fit is that we often get bogged down in the details over the exact dimensions and tolerances, based on a technical analysis. We forget that what we are trying to do is to design for the right feel.

7.7.7 Hardness

We also use the sense of touch to evaluate the hardness of an object. Not in a technical, calibrated manner, but in a more general way. We can easily discern whether something is rigid and unyielding (hard), or flexible and easy to deform (soft). Furthermore, most individuals can quickly ascertain whether they prefer the hardness of one material over another.

The hardness of a material can be quantified using a number of different test methods, many of which have their own specific means of testing and measurement. Most of these methods categorize materials based on where they fit within a certain range of measurements under a specific test protocol. This range is called a hardness scale. Since materials in the world range from very hard (diamond) to very soft (human skin), the test protocols vary a great deal, as do the scales. Hardness data are typically presented based on the test type, the test protocol, and the value within the scale. While there is overlap between some of the test methods and scales, there is no way to convert all hardness data into a true universal hardness system.

For thermoplastic materials, hardness is often described using the term durometer. A common test is the Shore durometer test, with some of the common test protocols being A, B, C, and D. A comparison of hardness data for some common materials are given in the graph below (Figure 7.22).

What this table does not reveal is the human response to hardness, the feel. However, the feel of material based on a given hardness is impossible

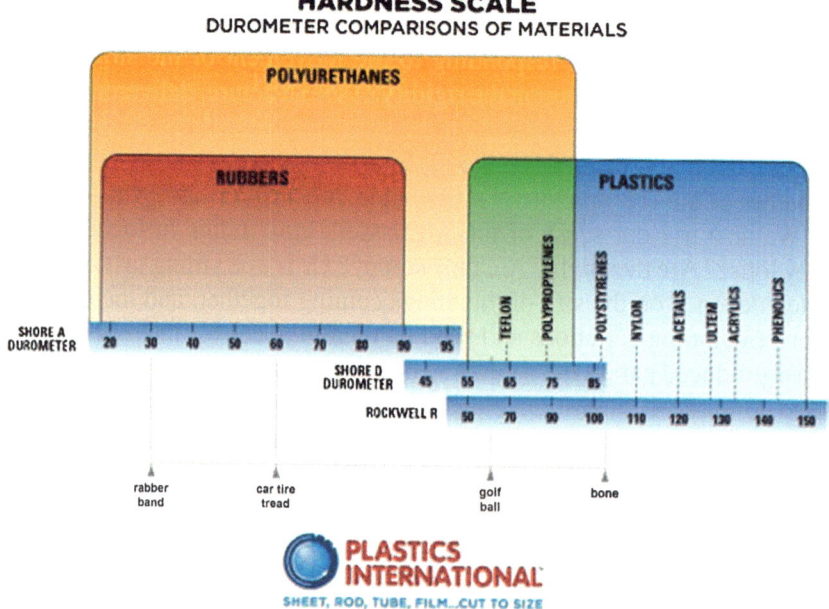

Figure 7.22 Comparison of hardness data.

7: Material Selection Based on Feel

to quantify. Our response to the material hardness will depend greatly on our expectations and desires for the application where the material is being used.

7.7.8 Texture

As discussed earlier, we often use our sense of touch to evaluate the texture of an object, that is, its surface roughness. And while our overall response to touching an object will depend on a number of variables, we rely a great deal on texture. We may use general terms to describe texture *(This feels smooth. This one is rougher than that one.)*, but we are quite capable of making very fine distinctions between different textures.

In woodworking, it is quite common to run one's fingers across the work piece as it is being sanded into its final surface finish. We can easily feel subtle imperfections that we cannot detect with the naked eye. With thermoplastics, while we may hand-sand or polish a surface finish into the part, more often than not the surface finish of a plastic part is created from one or more machining operations, either on the part itself, or on the mold used to make it. There is a wide range of textures that can be applied to thermoplastic parts, and the range is dependent on the manufacturing process. Most of these textures are selected based on visual criteria, although on occasion tactile criteria are used, as in *Will this texture provide a good grip?*

What I would like to encourage is an exploration of the question, *How will someone respond to the texture of this object when they touch it?*

Frequently, our response to texture is actually based on a combination of texture and other physical properties. This is especially true with synthetic fibers that come in contact with our skin. These fibers are sometimes woven into fabrics, other times they are simply clumped together, similar to how cotton fibers are clumped into a ball of cotton that you may use to swab your skin. Our sensation of touch involves not just the surface texture of the individual fibers, but the stiffness of each fiber, the material hardness, and how slippery it is.

7.7.9 Slipperiness

Slippery is a term used to describe how things behave when they slide against each other. We describe things that slide easily as being slippery, and things that do not as being sticky. Friction plays a role, as does adhesion. I would also argue that there is an element of liquidity involved.

We can easily detect the presence of liquid. Whether that is a drop of rain, a glass of water, or a gallon of gas. While we may smell something,

Figure 7.23 Swiss army slippery grip.

our primary means of detection is our sense of touch. It may be based on the way the liquid flows across our skin, or how it conducts vibrations, or the movement of the liquid itself. When we pick up clothes or grab a towel we know immediately whether they are dry, or wet, or slightly damp. How is that? Is it due to the weight of the water? A change in electrical or thermal conductivity? Or do we have some kind of moisture receptors in our skin? And we can also sense when a layer of film is present, either a water-based layer (mud, soapy film, cleaning solutions, etc.), or an oil-based layer (grease, grime, etc.). Then there are gelatins, like good old Jell-O® desserts. They are not really liquid, not really solid, but something in between.

In a similar manner, some people do not like to touch snakes. Something about the texture of the skin, and how it moves and slides. Other people get grossed out over frogs, or bugs, or other creepy-crawlies. We have terms like slimy, greasy, and gooey. And while all of this involves a combination of friction, adhesion, liquidity, and gelatinousness, I prefer to call it slipperiness. (I see a whole new book coming, *Goldilocks and the Seven Slipperies*.)

Selecting thermoplastic materials based on slipperiness is—for lack of a better word—a slippery task. Do you want it to feel as smooth as silk? (Silk fibers by the way, are not that smooth. The silky smooth feel involves more that just the surface roughness of the fibers.) When choosing a soft elastomer, do you want it to be squishy? How about slimy, gooey, and/or greasy? Do you want it to feel wet (even if it is not)? Do you want it to feel dry (even if it is wet)? Thermoplastic materials come in a wide range of slipperiness. While there is some overlap with tribology (the science of friction, lubricity, and wear), selecting materials based on how slippery they feel is whole different ball game (Figure 7.23).

7.7.10 *Material Selection Based on Touch*

We often speak about feeling a sense of warmth from having wood in our homes, from wood-beamed ceilings, hardwood floors, even wood furniture. I would argue that much of this is due to the sense of touch. We feel the deflections in a suspended wood floor as it flexes as we walk across it. We sense the low thermal conductivity as our bare feet come in contact with the wood, or as we rest our hands on a wood table top. Furthermore, the vibrations that are transmitted through wood feel different than vibrations that are transmitted through steel, concrete, and drywall.

The selection of thermoplastic materials based on touch is one of the most fascinating areas of plastics engineering. There are so many aspects to our sense of touch, and while some designers may consider some of these aspects, the vast majority have been overlooked by the plastics industry. However, many of us are already exploring avenues of touch and response without even realizing it. Designers and engineers in the sporting goods and automotive industries are always exploring new materials, including how we respond to them through our sense of touch. Medical-device designers have also been very focused on touch, although primarily from the perspective of the medical provider—not from a patient perspective. Regardless of the industry, there is much to learn about the human response to thermoplastic materials through the sense of touch.

7.7.11 Opportunities

For those of us who have designed hand-held electronic products, we know that at the end of every project, we are going to be tasked with coming up with a bunch of design solutions to address a plethora of issues regarding squeaks, clicks, rattles, and loose fits. This often involves a number of gaskets and elastomeric pads, each one custom made to address a specific problem. Why not just select a material at the beginning of the project that had a different balance of properties, with a more appropriate mix of stiffness, density, and damping factor? The end product would be cheaper to manufacture, and would provide a much better feel.

What if we could utilize thermoplastic materials to make a dental drill that instead of having an annoying high-pitched whine, had the soft whisper of a hummingbird on warm spring day?

What if Lady Gaga was a diabetic? What would her insulin pump look like? More importantly, what would it feel like? (Figure 7.24).

7.8 Smell

The sense of smell, also known as olfaction, consists of olfactory receptors in the nasal cavity, which then translate odor into electrical impulses which are transmitted by the olfactory nerve to the olfactory bulb in the

Figure 7.24 Have you seen my insulin pump?

brain. In humans, the olfactory bulb is located in a forward area of the brain, just below the nasal cavity. It is part of a set of brain structures known as the limbic system. In addition to olfaction, the limbic system processes memory, emotions, behavior, decision making, blood pressure, and adrenaline flow (the classic *fight* or *flight* response).

Odor is a generic term used to describe chemical compounds that are in a gaseous state. When these odors provide a pleasing response we often describe them using terms like scent, aroma, or fragrance. When they provide a less than pleasant response we describe them using terms like smell, stench, stink, and reek. Going back to our conversation about the language of feel, one of the challenges with olfaction is that we may use the word *smell* to describe not only the sense of olfaction, but also an odor that causes an unpleasant response.

7.8.1 Odor Detection

When we smell something, we are detecting airborne chemicals: volatile (meaning unstable, liable to disintegrate, or evaporate) molecules that are dispersed in the air. Objects that do not give off any sort of smell, like steel or stone, are nonvolatile solids that do not emit any free-floating molecules for the nose to detect. These airborne chemicals, also known as

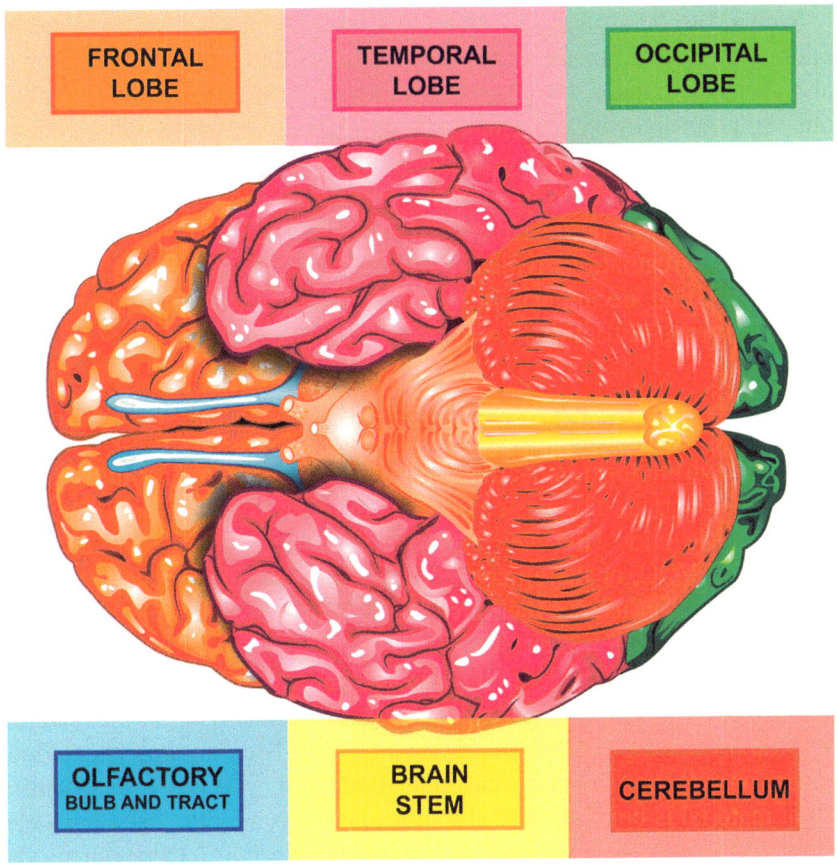

Figure 7.25 Human brain—underside view, showing olfactory bulb.

volatile organic compounds (VOC), originate from both man-made and naturally occurring materials. Our response to a given chemical will vary greatly depending on whether it is similar to geranyl acetate (the fragrance of a rose), or putrescine (the smell of rotting flesh), or skatole (the nasty smell in feces) (Figure 7.25).

As a species, humans are quite adept at detecting airborne chemicals, and just like with touch, we are capable of making very fine distinctions between different odors, or between different sets of odors. Our sensitivity to an odor will depend on the specific chemical involved. The term *odor detection threshold* is used to describe the minimum concentration of an airborne chemical that must exist for us to smell it. Thresholds for different chemicals are determined using a device known as an olfactometer. It is basically a device that can provide a quantitative answer to the question: *Exactly how badly does this thing stink?*

We tend to be attracted to odors that are good for us, and repelled by odors that are bad for us. In many cases odors that are disagreeable may also represent danger. The chemical we are detecting may adversely affect our skin, our digestive system, even our brain or central nervous system. Olfaction is often the first warning system for detecting something that should be avoided. Throughout the world, there are odor regulations, which address the permissible levels of airborne chemicals of varying kinds. These regulations typically address harmful chemicals, not necessarily those that are unpleasant. (As an example, I do not think there are any regulations restricting flower growers for fragrancy levels.)

7.8.2 Odor in Thermoplastics

In their virgin form, most thermoplastics are nonvolatile solids, or they have very low rates of emission of VOC. In other words, they usually have no detectible odor. However, a thermoplastic may often have additives which emit VOC, or there may be a trace amount of chemicals from the processing agents used in their production. The odor we detect is often due to the additives and/or the residue.

Most thermoplastics do emit some amount of VOC during processing. The amount and type of VOC will depend on the chemistry of the material and the processing conditions. If you have ever been on the factory floor where an injection molding machine is running, you can often identify what material is being molded through smell. (As a factory foreman once remarked to me, *I love the smell of nylon in the morning.*) Odor can also be generated from unintentional decomposition of the polymer as it is being processed (during molding, welding, etc.), or during combustion.

In most commercial applications of thermoplastics, odor is a nonfactor. Users have no expectation of smelling the material; they do not even consider the concept of a thermoplastic having an odor. There may be a comment about a plastic part that has been sitting in a closed container for a long period of time, but the odor quickly dissipates. Some users may comment about odor that is emitted during combustion, but rarely are thermoplastics used as a fuel source.

One area where odor in thermoplastics is being evaluated is in the automotive industry. Thermoplastics are used extensively in automobiles, and the vehicle cabin is an enclosed space. Almost all of us can remember being in a new car, and experiencing "that new car smell." People used to refer to this very distinctive smell with fondness, and later, with nostalgia once it was gone. It was a meaningful sensory experience, evoking a sense of accomplishment. Among other things, it communicated status—the

7: Material Selection Based on Feel 297

Figure 7.26 New car smell.

ability to buy a new car. To pay someone a compliment, you might sit down in the passenger seat and say, "*Ah, it still has that new car smell!*" After market suppliers even made air fresheners specifically intended to bring back that "new car smell"[13].

It was eventually discovered that the new car smell—which indicates the presence of VOC—actually has some negative health effects. That is because the odor is actually due to the off-gassing of various materials used in the fabrication of the vehicle interior. This includes not just the parts made of thermoplastics, but the solvents, dyes, and adhesives used in assembly, including the chemicals used in fabrication of those butter-soft leather seats. Many of these items would continue to release airborne chemicals long after the vehicle left the factory. In fact, there are often dozens or even hundreds of different compounds present, depending on the model of the vehicle [14]. And of course this issue is not just found in cars, but also in new airplanes, trucks, RVs, and other vehicles, practically anywhere that thermoplastics are used in an enclosed space. Some studies found that over time, in such enclosed spaces, these odors can cause adverse effects including dizziness, nausea, and headaches. In some cases, they can even cause serious internal damage, such as thyroid impairment and memory loss. Inhaling VOC in an enclosed space is, essentially, a form of "glue-sniffing" (Figure 7.26).

While the US still does not have regulations on VOC exposure in vehicles, many countries do, including Japan, which has some of the strictest standards and is also one of the world's principal auto producing countries [15]. With new international standards among automakers, many vehicle manufacturers have begun to pay more attention to producing low-VOC interiors. Some have turned to polyurethanes and polyolefins, which some say are less volatile, and are greener materials overall [16].

Material suppliers have also been working to address this issue. DuPont recently began manufacturing a low-emission acetal resin [17]. Celanese, another material supplier, has also developed a low-emission acetal copolymer called Hostaform XAP, that also meets original equipment manufacturer (OEM) performance requirements for low-VOC emissions. As technologies continue to develop, that familiar "new car smell" will gradually transform into a new ideal: "no car smell." Known as vehicle interior air quality (VIAQ), this is an important field requiring further research.

One of the interesting aspects of VIAQ has to do with the assessment of odor. There are international standards for the *measurement* of VOC in cabin interiors, but the industry is still struggling with odor identification. Odor identification is quite common in the perfume industry, where people with a highly developed sense of smell work as smell testers. They are able to identify specific odors (although in the perfume industry they use the term *scents*) with a high level of precision, and they understand the correlation between the scent and the chemical behind it. Unfortunately, the plastics industry is struggling with the correlation between the chemical compounds found in a vehicle interior and the odors they create. So as resin suppliers strive to create lower emission materials, they often do not know what compounds to focus on, or what processes to implement. A smell tester may describe an odor as smoky, which might involve dozens of different chemical compounds. As a resin supplier explores what compounds might be involved, and how to prevent their presence, the supplier is using the input of a smell tester and their sensitivity to a given odor. In a way, it is material development based on feel.

7.8.3 Human Response to Odor

Compared to other senses, human response to sensory input from olfaction is powerful and fast, often instantaneous. We may hear a piece of music and vaguely remember where we heard it before, or taste something and wonder, *Hmmm, what does this taste like?* But when we smell something that we have smelled before, there is often an instantaneous reaction to a long forgotten memory—whether that is the aroma of grandma's pumpkin pie on Thanksgiving Day, or the scent of your dog's wet hair from the early morning dew, or the fragrant cologne or perfume of your lover.

Furthermore, our physiological responses to odors are powerful and fast as well. The perfume industry has exploited this with great success, as have the food and beverage industries. But for whatever reason, the plastics industry has done very little to address human response to the odor of thermoplastic materials.

7.8.4 Material Selection Based on Smell

Thermoplastic material selection based on the sense of smell has primarily focused on the concept of selecting materials with no detectable odor. However, the concept that a thermoplastic material could have a noticeable and desirable odor has rarely been explored.

7.8.5 Opportunities

As we discussed in an earlier chapter, thermoplastics are polymers, created through the chemical linking of monomer molecules. While many of these materials were created intentionally, some were the result of serendipity, the unexpected consequence of an unplanned chemical reaction. Regardless of how they came into being, the raw materials used in their synthesis—the building blocks used to create them—are organic compounds.

What if we started with the premise that we wanted to create a thermoplastic that, as the polymer molecule degraded, it would break down into basic building blocks that smelled nice? We could create a thermoplastic material based on cinnamaldehyde, octyl acetate, or methyl anthranilate (the base chemicals of the odors found in cinnamon, citrus, and concord grapes, respectively). I do not know what kind of physical properties the resulting polymer might have, but using it might result in a delightful sensory experience—from the polymerization process to the final part, and hopefully to its end of life (Figure 7.27).

Figure 7.27 The aroma of freshly molded plastics.

7.9 Taste

The sense of taste, also known as gustation, consists of taste receptors on the tongue, which then translate chemical reactions into electrical impulses that are transmitted by a series of nerves to the brain stem. The brain stem is located in the rear of the brain, and connects the brain to the spinal cord. The brain stem also regulates the central nervous system. The main taste sensations are sweet, sour, salty, and bitter, along with a newer sensation described as umami (Japanese for savory). These taste sensations, combined with olfactory input and touch sensations in our mouth, give us the sensory input that we describe as flavor [18].

Compared to other senses, human response to sensory input from the sense of taste is pretty basic. Something either tastes good (in which case we have feelings of pleasure and satisfaction), or it does not. And while we may use our sense of taste to guide decisions about where to go for dinner, our survival instincts are rarely triggered by something we taste. The notable exception is our ability to detect toxic substances (which are often bitter), which is why we often have powerful, visceral reactions to bitter tasting substances.

The sense of taste is largely responsible for human response to chemicals released by certain materials. In 1931, DuPont chemist Arthur Fox accidentally released a cloud of a substance known as Phenylthiocarbamide or PTC. He could not smell or taste the cloud, but his colleague complained that it was bitter. Because he was curious, Fox then tested other people he knew, stumbling upon an important discovery in the field of genetics. The ability to taste certain substances is largely determined by genetic correlations, and as it turns out, ongoing studies have revealed that about 70% of people can taste PTC, with large variations in that percentage found in certain populations around the world [19].

Today scientists can actually manipulate taste sensations, by using chemical compounds that block or trigger certain receptors. Currently, major food producers use this science to make foods that taste fatty or sweet, but which are mysteriously low in calories. Have you ever looked at the nutrition facts for a bottle of nonfat nondairy creamer? Despite the fact that the substance is thick and creamy, and tastes fatty and sugary, the label says that it contains mostly zero everything—zero calories, zero fat, and zero nutrients.

There is much yet to be discovered about the sense of taste and it is likely that these discoveries will have implications for the future of the plastics industry. As the industry continues to develop new and better plastics, as well as plastics made out of renewable biomass sources, taste and smell will become important factors in their development.

7.9.1 *Material Selection Based on Taste*

The selection of thermoplastic materials based on human response to how they taste is not something most of us think about. If anything, we select thermoplastic materials based on the specific requirement that they have no effect on taste. And while some people may make flippant comments about plastics and bad taste, this is mostly a social commentary and has nothing to do with gustation.

However, we often use thermoplastics to store and serve food, beverages, and medications, and there are thousands of products that we put in our mouths that were probably not intended to be placed there. There are also countless other products, such as thermometers, pacifiers, and dental instruments, which are specifically designed to be placed directly in the mouth. But in all of these products, taste is rarely considered. The sense of touch may be considered, either via the use of texture or the density of the material or its hardness (and/or softness). We may also consider the thermal conductivity of a material or perhaps even its electrical conductivity. But taste? Rarely (Figure 7.28).

Yet at some point in time, most us have chewed on something made out of plastic, perhaps without thinking about it. And while the designer of that product did not intend for it to be chewed on, it happens. Question: *Knowing this, what material should they have selected, and why?* (The discussion is left to the reader.)

In some materials there are additives known as plasticizers, which are used to make the material softer and more ductile. Some plasticizers are phthalates, which are esters of phthalic acid. The use of phthalates is a controversial topic and there is considerable debate about their impact on

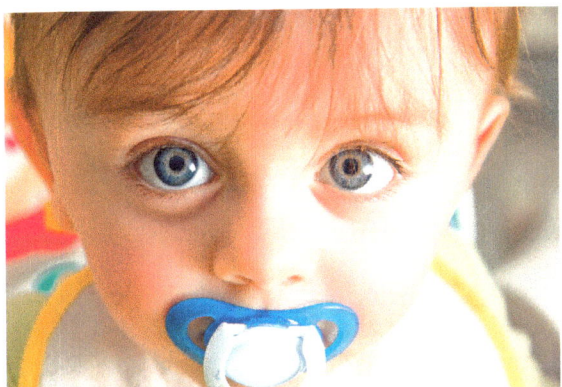

Figure 7.28 The taste of plastic.

Figure 7.29 Edible plastics.

health and safety. Even so, the debate is not about taste. Perhaps the industry may soon develop flavor additives for thermoplastics. It would give a whole new meaning to the phrase, *Tastes Great, Less Filling*.

7.9.2 Opportunities

What if children's toys were made of materials that looked good and performed well, and felt wonderful to touch, but tasted horrendous? Any child that puts that toy in its mouth would immediately spit it out (and never do it again).

What if we could embed a nasty flavor in our most commonly discarded plastics—the materials that often end up in the stomachs of birds and fish and other wildlife—so that in the event they end up in the environment, animals would avoid them like the plague?

What if we could make a thermoplastic material that was not only biodegradable, but also edible and digestible? We could use it for food packaging, so not only could you have your cake, you could eat the package too. We could add a flavor to it and the whole event becomes a dining experience (Figure 7.29).

7.10 A Methodology

The methodology for selecting thermoplastic materials based on feel is not that much different than methodologies based on performance and/or cost. It begins with a product specification, followed by a preliminary exploration of material candidates. These materials are then evaluated in greater detail, some prototypes are fabricated, the design is

7: MATERIAL SELECTION BASED ON FEEL 303

revised, and a final material selection is made. While there are different tasks involved when selecting materials based on feel, the primary difference from other methods is that it is a subjective approach.

7.10.1 Infrastructure

A critical component of this process is the infrastructure within the company. Most companies have product development teams, perhaps involving R&D, engineering, industrial design, marketing, and manufacturing. They also have facilities and equipment for design and development, and for testing and verification. However, most companies have not allocated any resources toward the assessment of feel. While many companies have committed resources to industrial design, their efforts are often focused on form and visual appearance. Although this is often a critical factor in the success of a product, feel involves more than just form and appearance.

On a management level, resources need to be allocated for the material selection process itself. This involves getting the right people and the right equipment, and scheduling time for evaluation. While finding the right people is critical in all aspects of business, it also applies to those involved in assessing feel, or as I often describe them, *The Feel Team*.

The feel team is not the entire development team. Instead, it is a dedicated group of individuals with exceptional abilities in the feel-related aspects of design. We all know these individuals. Sometimes they are designers, but they are often engineers or marketing specialists. They are people with a knack for the subjective, nontangible elements of feel. Perhaps they are also artists or musicians or photographers with sensitivity to color, sound, or light. Dancers and athletes often have a highly developed sense of touch, as do massage therapists. As we described earlier, smell testers have a highly developed sense of smell. The make up of your team will depend on the industry you are in, and what aspects of feel are most applicable.

It is important that each member of the feel team has good communication skills and substantial people skills as well. An individual could have all the sensitivity in the world, but their input is useless if they cannot communicate what they are sensing to the rest of the team. They also need to be committed to the process. Nothing is worse than having a naysayer sit in the room who rolls their eyes every time someone else says something they disagree with. Lastly, make it a diverse team. Remember, feel is subjective, and the more diverse the evaluators, the better the collective evaluation.

A clear decision-making process must be established. This is particularly important for the final decision on material selection. Since feel is subjective (and rarely unanimous) there is going to come a point where a decision has to be made. Identify the key decision-maker up front. This is the person who will be responsible for the final choice of material. It might be the project leader, a senior engineer, a manager, or even the company owner.

Finally, accept the basic premise that perfect feel does not exist. To paraphrase Abraham Lincoln, "You can improve the feel for some of the people all the time, and for all the people some of the time, but you cannot improve the feel for all the people all the time [20]."

7.10.2 The Process

Once you have the infrastructure in place—including a feel team—the material selection process can begin. Below is a list of some suggested steps to help guide the process.

Feel Specification—Write a feel specification. Describe, as best as you can, what the desired feel is (and what it is not). Prioritize the various elements of feel. Is color most important? Or is it the gloss level? Or is it the sense of touch? Is sound more important than vibration? Or is vibration more important than sound? Consider the question—"Would I like this part to feel like something I have felt before?" (because I had a favorable response to the way that thing felt). Treat this feel specification as you would any other project specification.

Material Recommendations—Get recommendations on material candidates. Seek out a wide variety of sources: end users to provide insight and feedback on their "feel" toward existing products and/or what you're designing, "lessons learned" from prior design teams and competitive evaluations, other insightful sources.

Swatches—Gather lots and lots of swatches: test plaques, color chips, other parts, samples of commercial products, prototypes from earlier projects, etc. The collection should be based on your feel priorities and your starting point.

Team Evaluation—Gather the team and evaluate the swatches. Explore what feels good, and what does not. Explore comments and ideas about why the swatches feel the way they do. Use the Goldilocks Equation in your evaluation efforts. Note: At this step, it is often helpful to start writing things down.

Material Exploration—Based on the team evaluation, select materials for further evaluation. Do additional research on the materials: get data

7: Material Selection Based on Feel

sheets, design guides, case histories, explore pricing and availability, maybe even get additional samples. In your exploration, consider the following: What aspects of these materials are affecting the feel? Is it stiffness? Hardness? Density? One specific property? Or a combination of properties?

Make Prototypes—One of my favorite mantras in product development is: *Prototype Early, Prototype Often*. It may sound old-school, but even with today's technology there is nothing like having a physical prototype in front of you. When it comes to evaluating feel, physical prototypes are mandatory. You will learn things that no amount of design research or analysis can predict. The exact number of prototypes you should make depends on a number of variables, including the complexity of the design, the costs involved in prototyping, and your familiarity with the materials selected (as well as how you feel).

Iterate—As you evaluate materials and prototypes, be prepared to revisit various steps in the selection process. How often you revisit these steps will depend on the industry, the application, the specific aspect of feel that is under evaluation, and the evaluation process.

Note: the word, iterate comes from the Latin word *iteratus*, which means to do a second time, revise and renew, repeat. Far too often, when it comes to material selection based on feel, or product development in general, companies often confuse the meaning of iterate with Albert Einstein's definition of the word insanity: *doing the same thing over and over again and expecting different results* [21]. I use the word iterate to mean *to do again*, with the intent of rehearsing, refining, and perfecting. An analogy would be the classic joke: Q. "Pardon me sir, but how do you get to Carnegie Hall?" A. *Practice, practice, practice* [22].

Make a Decision—After all the exploration is complete, and all the prototypes have been made and evaluated, there will come a time when you need to make decisions. You may not have all the test results, all of the user feedback, or all of the cost data. However, decisions on the materials will need to be made, and they must be made by the key decision-maker.

Final Prototype—Prior to giving approval on the final design and materials that will be used in mass production, make one final prototype that incorporates all of the findings in your selection process. This prototype (or prototypes) should incorporate parts that are similar to—if not identical to—the size and shape of production parts and are fabricated from materials that are similar to—if not identical—to the production materials. This final prototype represents the look and feel of the production system, before production tools and parts are available. It provides one last chance to confirm all aspects of feel before production tools are ready. Note: This is an optional step, but one that is highly advised.

While the above steps are presented sequentially, material selection is rarely a linear, sequential process. More often that not, material selection is an iterative process, involving both trial-and-error and trial-and-success. As such, it is expected, and often advisable, to move back and forth between steps, or to address multiple steps concurrently (Figure 7.30).

7.10.3 Making It Work

Material selection based on feel is a simple process. While it frequently relies on subjective evaluation, at its core it involves comparison: Does material A provide a more desirable response than material B? As with any methodology, the set of best practices is constantly evolving, and may vary depending on the industry and the company.

Unfortunately, many companies ignore feel completely. They think that dealing with the feel of a material is a waste of time and money, and that resources could be better spent on other issues. So they avoid dealing with any issues involving feel or they cover them up using a band-aids and duct-tape approach. Even companies that spend megabucks on industrial design often neglect the feel of the materials they use.

Years ago (in a galaxy far, far away), while I was working in the cell phone business, we were dealing with heat issues on a certain phone. The damn phone was literally too hot to handle. The mechanics team was tasked with evaluating the heat that was generated within the device, identifying the root causes, and coming up with proposed solutions. One of the threads of exploration involved the concept of comfort, as in: *What is a comfortable temperature for the housing of a hand-held device?*

As the team started exploring, it became evident that the notion of comfortable temperature depended on the material. A metal case conducted heat differently than a plastic case, and at a certain temperature it "felt" too hot, while a plastic case did not feel too hot until a higher temperature. Similar observations were reported for different applications and different

7: Material Selection Based on Feel

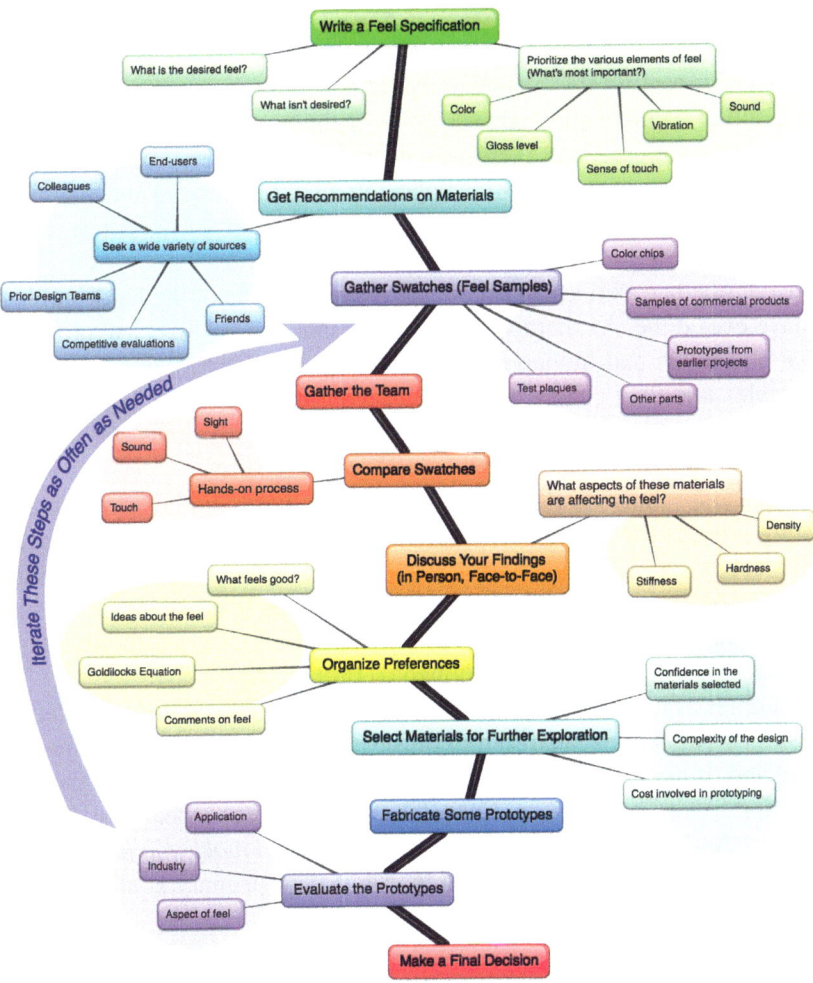

Figure 7.30 A flow chart for selecting materials based on feel.

materials, ceramics versus metals, plastics versus metals, stone versus steel, etc. It was very much a feel thing. The technology behind this rather was interesting (and this experience was the foundation of the earlier section on temperature), but what was more interesting was the reaction of the industrial design team and the overall response of the company. They did not care about the technology, they did not want to focus any resources on the problem, they wanted the phone to look a certain way and they simply wanted the mechanics team to fix the heat problem. Needless to say, this company is no longer making cell phones.

7.11 A Final Word about Feel

As I was writing this chapter, I heard news that a close friend had suffered a bad accident and had been taken to the hospital. He was in a coma, and on life support, and there was no hope of recovery. His family had made a decision to turn off life support in a couple of days and wanted to offer his friends a chance to see him one last time and say good-bye. I went to see him the next day.

He looked peaceful. But saying good-bye was one of the hardest things I have ever done. As therapy, I dove back into my writing. I wrote about heat and touch and human response and product performance. I worked on tables and flow charts and compelling arguments. But I kept thinking about my friend, lying in a quiet room, surrounded by tubes and blinking lights.

The funny thing is, I could probably name the material and manufacturing process used for every part on every product in that room, ranging from touch screens to tubing, tissues to toilet seats, even the fabrics and fibers used on his hospital gown. Thermoplastics everywhere. It was not until some time later that a question came to mind. "What thermoplastic should I specify, in order to help mend a broken heart?"

In times of distress, we often rely on habits and rituals that are intended to soothe our senses. We listen to music. We sing and we dance. We have food and drink to nurture our bodies and to comfort us. Food for the soul. Comfort food. Chicken soup. Cake. Ice cream. Flowers—beautiful, fragrant flowers. And other therapies. *Here, take this, it will make you feel better.*

But what is the proper material to use, if your primary goal is to mend a broken heart?

I do not really have an answer.

References

[1] Esalen Institute. http://esalen.org.
[2] Ivan Pavlov, Wikipedia: The Free Encyclopedia. http://en.wikipedia.org/wiki/Ivan_Pavlov, actually it was not the ringing of a bell, but other stimuli.
[3] Loosely Attributed to Gloria Steinem, American Feminist. Also Often Quoted as: A Woman Needs a Man like a Fish Needs a bicycle.
[4] On the Soul (Latin *De Anima*), Book II, Chapters 5–11. Aristotle.
[5] Jennifer Kahn, Amazing Facts about Your Senses. Parade's Community Table, July 29, 2012.

[6] Faber Birren, Color & Human Response: Aspects of Light and Color Bearing on the Reactions of Living Things and the Welfare of Human Beings, 1984.

[7] [a] Pam Belluck, Reinvent Wheel? Blue Room. Delusion a Bomb? Red Room, The New York Times, February 6, 2009.
[b] Robert Roy Britt, Red vs. Blue: Why Necktie Colors Matter, LiveScience, February 6, 2009.
[c]Elliot, Andrew J., Markus A. Maier, Arlen C. Moller, Ron Friedman, and Jörg Meinhardt. Color and psychological functioning: the effect of red on performance attainment. J. Exp. Psychol. 136(1):154-168.
[d]Elliot, Andrew J., Markus A. Maier, Martin J. Binser, Ron Friedman, and ReinhardPekrun. The effect of red on avoidance in achievement contexts. personality and social psychology Bulletin 35(3):365–375.

[8] K. Palana, Between the Lines: How Politicians Use Color Psychology to Win Your Votes, The Lamp, February 28, 2012.

[9] Hugo Fastl, Eberhard Wicker, Psychoacoustics: Facts and Models, Spring Series, 2007.

[10] [a] O. Sacks, Musicophilia: Tales of Music and the Brain, Vintage, 2008.
[b] Floyd Toole, Sound Reproduction: The Acoustics and Psychoacoustics of Loudspeakers and Rooms, Focal Press, 2008.

[11] Rick Chillot, The Power of Touch, Psychology Today, March 11, 2013.

[12] John Roach, Elephants May 'Talk' via Vibrations, National Geographic News, July 8, 2002.

[13] R. Jaslow, New Car Smell Is Toxic, Study Says: Which Cars Are Worst? CBS News, February 15, 2012.

[14] [a] T. Brockmann, What Causes a New Car Smell? WiseGeek, September 15, 2014.
[b] Ryan Jaslow, New Car Smell Is Toxic, Study Says: Which Cars Are Worst? CBS News, February 15, 2012.

[15] [a] Stephen Moore, Low VOC Emission Standards for Vehicle Interiors Propel Polyacetal Development, Plastics Today, May 18, 2014.
[b] Mark Polster, Vehicle Interior VOC Measurement. Ford Motor Company Environmental Engineer Report, Society of Automotive Engineers VOC Committee, 2014.

[16] Wendy Koch, Most and Least Toxic Cars? USA Today, February 17, 2012.

[17] DuPont, DuPont (TM) Delrin (R) Extends Scope of Use of Plastics in Vehicle Interiors, 2014.

[18] Tanya Lewis, The Bittersweet Truth about How Taste Works, LiveScience, March 6, 2013.
[19] Steven Wallace, Phenylthiocarbamide, Chemistry World, October 13, 2013.
[20] A Paraphrase of a Quote Attributed to Abraham Lincoln, 16th President of the United States, 1809–1865. The Actual Quote: "You Can Fool Some of the People All of the Time, and All of the People Some of the Time, but You Cannot Fool All the People All the Time."
[21] Attributed to Albert Einstein, United States (German Born) Physicist, 1879–1955.
[22] This Quote/Joke Has Been Attributed to a Number of Sources, Including Arthur Rubenstein, United States pianist, 1887–1982, and Jack Benny, United States comedian, 1894–1974.

Further Reading

Francis M. Adams, Charles E. Osgood, A cross-cultural study of the affective meanings of color, Journal of Cross-Cultural Psychology 4 (2) (1973) 135–156.

Augustine Hope, Margaret Walch, The Color Compendium, Van Norstrand Reinhold, 1990.

Itten, Johannes. The Elements of Color.

J.L. Morton, Color Logic, Colorcom, 2008.

8 Bringing It All Together

In our journey through the world of thermoplastics, we have touched on a number of subjects, to help guide the reader in their use of these materials. As we have discussed, the science behind thermoplastic materials is highly specialized, and it often requires specialized expertise to process them effectively. But the fact remains that thermoplastics are materials that are used to make things. Just as humans have used wood, stone, and metal to make things, we are now using thermoplastics.

However, unlike wood, stone, and metal—which humans have been working with for thousands of years—plastics have only been around for a few generations, and just in the last few decades that they have come into widespread use. For most of us, our education in design, and/or engineering, architecture, medicine, etc., may not have included hands-on training in the use of thermoplastic materials, or in selecting the right material and the right process in order to make a quality part at the right cost.

In this chapter we will bring together all our previous discussions, and provide a summary of the material selection process. We will review the basic methodology for selecting a thermoplastic material, which culminates in a final material specification. Next, we will review the plastics supply chain, the supplier relationships involved, and how to troubleshoot problems. Finally, we will provide a list of resources for further investigation.

8.1 Material Selection

Throughout this book we have emphasized that effective material selection is about finding one or more suitable materials that—in combination with an effective design, proper processing, and eventual integration into a final system—results in a product that satisfies the needs of an end user.

When we think of a well-designed product, we often think of its overall shape and form, its ease of use, and how well it performs. However, most well-designed products are also products that have been manufactured to meet or exceed industry standards. This is only possible when the component parts themselves are properly designed and manufactured. The foundation of good manufacturing begins not with proper processing controls, but by having a robust part design with the right material using the right manufacturing process. These three items—part design, material selection, and manufacturing process—are the cornerstones of good manufacturing. Of these three, material selection is often the most troublesome, especially when it comes to selecting a thermoplastic material. One could even conclude that effective material selection is the foundation of good design.

Let us review the basic steps involved in selecting a thermoplastic material.

8.1.1 Step One: Establish Key Criteria

Early in the design process, we need to establish the key criteria for each part. The criteria may include performance requirements, cost targets (for both part costs and tooling costs, perhaps even for equipment costs as well), and feel issues. (While feel requirements are often overlooked, they sometimes are the most important criteria.)

8.1.2 Step Two: Select the Manufacturing Process

Next we select a suitable manufacturing process. The variables to consider in our selection include part size, part complexity, product volumes, and allowable tooling costs. Of these, production volumes are usually the driving factor, followed closely by tooling budgets.

8.1.3 Step Three: Develop a Short List of Material Candidates

Based on our key criteria, we develop a list of material candidates. This list should be grouped by chemical family (i.e., the polypropylene family, the nylon family, etc.), as well as by primary additives that may be needed (glass reinforcement, tougheners, heat stabilizers, flame retardants, etc.). Potential suppliers can also be identified at this time.

8.1.4 Step Four: Evaluate the Data

Next we begin the process of evaluating data. This includes the material data for mechanical, chemical, and/or electrical properties, as well as cost estimates, structural analysis, and supply chain explorations. It may also include an evaluation of color, sound, vibration, thermal conductivity, or other aspects of feel.

These evaluations may be simple and straightforward, or they may be complex. The level of effort involved will vary, and the time that is required can range from a few days to several months, depending on the complexity.

8.1.5 Step Five: Development

The next step in material selection occurs during the development phase. This phase is an iterative phase and involves detailed design and engineering, prototyping, testing, and often modifying and re-testing. This phase is usually the longest phase in terms of time. During this phase the short list of materials will get even shorter, as one or perhaps a few materials are proven to be most effective.

Development can occur in either a research & development environment (R&D), or a product development environment (PD or NPD). In an R&D environment, the focus is on proving out a technology, and the deliverable at the end of the phase is knowledge, or an approved procedure, or a material choice that is proven to be effective. In an NPD environment, the output is a specific product.

8.1.6 Step Six: Select the Material

The last step is to finalize the material selection. This selection should be obvious, with a material that meets all of the established criteria, at the most effective cost. There should not be any guessing, or any crossing of the fingers and hoping for the best. It should be a fact-based decision, based on a robust evaluation, with reliable, proven data.

If a material choice is not obvious, or all of the key criteria have not been met, the material choice should not be finalized. Some of the earlier steps may need to be repeated. Additional evaluation may be needed, or more development work. The criteria may even need to be modified. In an NPD environment, this may cause project delays. However, it is often advantageous to accept that delay now, and prove out the right material, rather making a poor choice—one that may cause more severe delays (and potentially much higher costs) in the future.

8.2 Material Specification

Once the material has been selected, a material specification should be written. Many companies have formal material specification documents, which describe the material, some of the required properties and sometimes even the test methods and procedures used to verify those properties.

These documents will usually also reference specific products from specific suppliers in an "approved materials" section.

Many international standards organizations have procedures for writing a material specification. One such organization is ASTM International (which was previously known as the American Society for Testing and Materials). By following these procedures one can write a material specification in a clear and comprehensive manner, and it can be read and understood by anyone who has access to the procedure document itself.

At its most basic form, a material specification should describe the material in a clear, unambiguous manner. While this can be done by referencing a product code from a resin supplier, it can also be done in a narrative form, by describing the chemical family, the common name, the additives, etc. An example is given below. It is highly advisable to provide a reference to an international standard and require the material supplier to verify that the material meets that standard. The advantage of a narrative specification is that it provides a reader with descriptive information, which may not be immediately obvious from only a product code.

Chemical name: polyamide

Generic name: nylon

Common name: nylon

Type: 6/6 homopolymer

Reinforcement: 30% glass fiber

Required additives: heat stabilizer

Permissible additives: internal lubricant

Color: natural color (no colorant)

Certified to: ASTM D4066 PA012G35

8.2.1 Approved Suppliers

A common practice in material specifications is to provide a list of one or more approved material suppliers. This is usually done in conjunction with the supplier product code for the material that has been specified. There are many reasons for doing this. It provides a confirmation that the material you specified will be used, exactly as you specified. This supplier may have a reputation for quality, reliability, and global availability. It also provides the foundation for a long-term business relationship. You are rewarding the resin supplier for the technical support they provided

8: Bringing It All Together

in the evaluation and development phases, and entering into a supply partnership.

One disadvantage to this practice is that the molder or part supplier may not have a relationship with that supplier, or they may only use a limited amount of material from them. As a result, they may not have a favorable supply contract, and there may be higher pricing. This can be avoided by working with the molder and/or part supplier during the evaluation and development phases to ensure that they have a favorable relationship with all of the suppliers of the materials that were on your short list.

8.3 The Plastics Supply Chain

In the manufacturing of plastic parts, the supply chain is not a linear relationship; it is more of a network. It begins with the resin suppliers, who sell raw materials to converters. The converters buy the raw material, as well as processing machinery and tools, and convert the resin into molded parts, and then sell the molded parts to the product manufacturer (Figure 8.1).

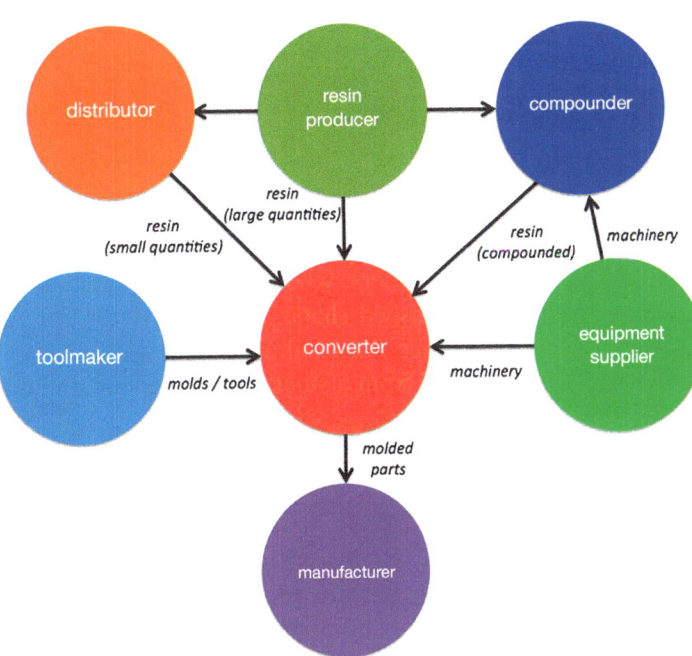

Figure 8.1 The plastics supply chain.

8.3.1 Resin Suppliers

The primary source of resin supply is the resin producer. The producer may be a small, specialty producer, or a major international producer with a large portfolio of resins. Regardless of their size, the producers are the creators of the resin. They may sell the resin to a molder or manufacturer directly, or they may sell it through a distributor. While distributors will have knowledge in the processing and use of the resin, the primary expertise, and the source of most plastics data, are with the resin producer.

8.3.2 Compounders

Plastic compounders are companies that purchase bulk material from the resin producers, and then formulate custom compounds through the mixing and blending of base polymers with various additives. While some resin producers do some basic compounding, it is typically on a limited scale for a few specific resins (usually only standard materials are produced in high volume). Most compounders offer a wide variety of resins in a number of different versions—in high-volume or low-volume production—and are also capable of formulating specialized compounds on demand.

Collectively, resin suppliers, distributors, and compounders are the sellers of plastics as raw materials.

8.3.3 Converters

The next link in the supply chain is the converters. These are molders, extruders, and stock shape suppliers, who buy plastic materials, and then melt and form them into a finished shape. Some converters will act as a supplier for a second converter. An example would be a thermoformer, who would buy sheet material from a stock shape supplier, and then convert it into a finished part.

8.3.4 Equipment Suppliers

Equipment suppliers provide equipment to plastic processors, primarily to converters, but also to compounders. (Resin producers typically build their own equipment.) Common equipment includes molding machines, dryers, feeders, handling and transport machinery, and robotics.

8.3.5 Toolmakers

Toolmakers are a specialized subset of the supply chain. They provide customized molds and tools that are used by the converters to create plastic parts. Toolmakers may be employed by converters directly, or they may be a supplier to the converter. The molds and tools they make are typically designed so that they can be used in most standard processing equipment, but they may also be designed to fit only in a specific machine.

8.3.6 Product Manufacturers

Product manufacturers are the companies that are producing the final product. They are the buyers of the parts that were produced by the converters. Some product manufacturers may have molders and converters that are internal to the company, or are dedicated to producing parts only for them (this is known as a *captive molder*). Some product manufacturers may have their product assembled by another company on a contractual basis (these companies are known as *contract manufacturers*). Also, some manufacturers may use intermediate suppliers who specialize in building specific subsystems. These system suppliers buy or fabricate the component parts, build the subsystem, and then deliver it for final assembly. They may use captive molders as well.

The product manufacturer is the final link in the chain. While the ultimate customer is the end user of the finished product, in the supply chain network the product manufacturer is the primary customer. The product manufacturer is responsible for the performance of the product, and all of the parts that go into it. As such, they are also responsible for material selection. While all of the participants in the plastics supply chain can provide recommendations on materials, the final decision on material selection should be made by the product manufacturer.

8.4 Industry Infrastructure

In addition to the supply chain, the plastics industry has a well-developed infrastructure. This includes trade organizations, professional societies, academic institutions, research centers, and an extensive network of service providers.

8.4.1 Trade Organizations

Trade organizations in the plastics industry include umbrella organizations, such as the Society of the Plastics Industry, and groups with a specific focus, such as the American Mold Builders Association, or the Association of Rotational Molders. There are also professional societies, such as the Society of Plastics Engineers or the Polymer Chemistry Division of the American Chemical Society. Many of these organizations sponsor trade shows, which may be regional, national, or global in scope.

There are also a number of standards organizations, which provide standards in testing, measurement, and materials science (Figure 8.2).

8.4.2 Education

Most major universities throughout the world offer mechanical engineering programs, and more and more have programs in materials science. Unfortunately, most of these programs have limited exposure to the world of

Figure 8.2 Trade organizations in the plastics industry.

thermoplastics. In the United States, there are only about a dozen universities which provide degree programs specifically on plastics. These programs are sometimes oriented to polymer engineering, other times to polymer science. A few are focused exclusively on plastics engineering, which is dedicated to the design, development, and manufacturing of plastic products. A listing of some of these programs is given in the resources section.

In addition to degree programs, there are extensive educational resources available. Most major resin suppliers offer design guides, which are often available free of charge, and there are technical books and trade magazines.

8.4.3 Information Providers

There are a number of sources which can provide detailed information about plastics. This information may involve market research, material properties, test methods, or industry applications. Many suppliers offer some of this information free of charge, and many of the organizations described above have technical journals, databases, and service directories available to their members. There are also online tools such as search engines and Wikipedia, which can provide basic information with a few clicks. There are also paid information sources, including some online databases and libraries.

8.4.4 Plastics Testing

Testing is a critical aspect of plastics performance, and there is an extensive network of testing services in the plastics industry. This includes manufacturers of test equipment, independent testing labs, and industry organizations that develop test standards. Test equipment manufacturers supply equipment to all of the participants in the supply chain, and may also offer test services on a contracted basis. Most resin suppliers have extensive in-house testing, primarily for material characterization, but they may also offer test services for customer applications. Independent testing labs typically offer services on a contracted basis and are often considered the private investigators of the industry.

8.4.5 Service Providers

The industry infrastructure also includes a supplier base that provides goods and services in the plastics supply chain in various ways. Some providers focus on the business side of the plastics industry. They may provide expertise on the supply chain (from resin production to manufactured part), or in pricing, or in legislative activities. Other providers focus on the technology side and provide services in design, engineering, analysis, and/or manufacturing.

One important technology issue is materials technology itself. This includes polymer chemistry, rheology, and material characterization. The providers may be specialists in a given chemical family, such as PVC or nylon, or in a specific polymerization process, or they may be more generalized.

Another technology area involves polymer processing, including material handling, drying, injection molding, tool maintenance, plastics painting, or plastics decorating. And there are those who work in plastics engineering, including application engineers, structural analysis experts, design specialists, and even design wizards.

Service providers may be small firms, large international companies, or independent consultants. They participate in the supply chain in different ways, depending on their area of expertise.

8.5 Working with Suppliers

As with all businesses, suppliers play a critical role in the success of a manufacturing company. Good suppliers can help you succeed, while bad suppliers can put you out of business. In plastics manufacturing, there is a diverse network of suppliers in the plastic supply chain and the industry infrastructure. Finding the right supplier for your needs is often as important as selecting the right material.

8.5.1 Determining Capabilities

At one time or another, most of us have had something custom made. In the process, we probably got involved with the fabricator of that item. There may been questions like: *Can you make this thing for me, exactly the way I am describing it? Are you sure? Have you ever made anything like this before? Can you show me?*

Finding a supplier for a manufacturing operation involves asking similar types of questions. From there the conversation may shift to a review of the supplier's machinery and manufacturing processes, their experience with industry standards and protocols, record keeping, quality control, etc. While there are many ways to evaluate all of this, the ultimate goal is to confirm, *yes, this supplier is capable of making this part/system/product.*

In regards to thermoplastics, it is important to confirm that a supplier has experience with that exact material, making parts that are similar in size, which meet similar performance criteria. Do they mold glass reinforced nylon 6 in hand size parts that must withstand high impact? Do they use polycarbonate to mold small optical lenses? Do they thermoform ABS sheet into structural housings with high gloss requirements? Regardless of whether the material is amorphous or semicrystalline, toughened, or flame retarded, or glass reinforced, they should have experience with the material.

8.5.2 Determining the Right Fit

Once you have established the supplier's technical capabilities, the next step is to discuss the elements of quality, timing, and price. As the old sourcing joke goes, you get to pick two. You can get high quality at a good price if you are not in a hurry, but if you want high quality and you want it quickly, it is going to cost more. While price is always a consideration, it may not be the most important factor.

Equally important is the size of the vendor. Are they capable of producing the part in the quantities you want? Are they too small and do not have the production capacity? Would you need to find an additional supplier as well? Perhaps it would be better to find a larger vendor with the capacity to deliver the quantity you need. Although, if that supplier is too large, fulfilling your requirements may not be their top priority, or they may not be willing to provide the level of service or responsiveness that a smaller supplier might be able to provide.

8.5.3 Project Participation

Once you have confirmed capability and fit, a simple next step is to move forward with a project. Some companies will enter into a business relationship with a supplier only after an extensive evaluation process. Some companies may use a small project as part of the evaluation process. That project will often have a well-defined scope of work, to be delivered over a certain period of time, at a specified price. This kind of project can often be beneficial to both parties. When the work at hand gets completed, capabilities and competencies can be evaluated in greater detail, and the working relationship can also be explored.

8.5.4 Managing the Relationship

The relationship a manufacturer has with its suppliers is critical. This involves more than just having a good fit between customer needs and supplier capabilities. The customer–supplier relationship, or perhaps we should say relationships, can have far-reaching affects, not just on the manufacturing floor, but in engineering, accounting, sales, even corporate branding.

A good relationship takes time to develop and takes effort to maintain. While we may sometimes think of suppliers as providers of goods and/or services, they are a business partner as well. There is often a direct connection between the supplier relationship and the success of the final product.

8.5.5 Communication

As in all relationships, it is important to have effective communication with your suppliers. Most companies go to great lengths to describe the financial terms of the supplier relationship, using pricing contracts, supplier agreements, etc. These documents usually describe functional requirements, delivery schedules, along with pricing and payment terms. There may also be additional technical documents, such as technical reports, design briefs, product specifications, and engineering drawings. There may also be informal communications, including email exchanges, phone conversations, and face-to-face meetings.

In all of these interactions, the goal should be to *communicate*. Communication is a participatory event, meaning it is in activity that one has to *take part in*. It involves much more than just releasing documents and writing memos. Effective communication also requires hearing, listening, and understanding. Effective communication is the foundation of a good supplier relationship.

8.6 Troubleshooting

In all businesses, and in all aspects of life, there are times when we encounter problems. Sometimes these problems are small, temporary issues, which are really nothing more than an annoyance. We quickly fix them, or maybe they fix themselves. But in the world of manufacturing there are times when a small problem can quickly escalate into a big problem, one that results in a safety hazard, or a product recall, or even a loss of life.

8.6.1 Assemble the Team

The first step in troubleshooting is to assemble a team. It might be a small team, or it might be a large team, depending on the problem. It may consist of people familiar with the project, or people from outside the project, again, depending on the problem. The important thing is to have people with the right skills, not just technical skills, but problem solving skills.

Years ago, I had a boss who had a tendency to speak his mind, often rather forcefully. During one project review, the engineering team had been describing some of the problems we had been working on, when he interrupted us. *I don't need problem identifiers. I need problem solvers. Can you get this done?*

It was an interesting comment, and it highlights a critical aspect of troubleshooting. When it comes to problem solving, some people are good at it, and some people are not. I am sure there have been studies on this. Perhaps it is a mix of critical thinking skills, problem-solving methodology, creativity, and a willingness to take risks. Perhaps it is based on common sense. Whatever the reason, your troubleshooting team needs to have people who know how to solve problems.

8.6.2 Identify the Real Problem

Before you can solve the problem, you first need to identify it. As engineers, we are often presented with problems, and our training is to jump into analysis mode. This can be counter-productive, since there are often things going on that are not immediately obvious. If we do not take time to investigate and understand what is happening, we may be solving the wrong problem.

So, what is the problem? *The problem is, these parts are breaking!* Really? Duh? That is not identifying the problem, but simply making a statement of the obvious. A good investigator knows you have to ask questions, often lots of questions.

The world is filled with stories about business successes that were the result of a simple question. These stories usually involve lots of feel good moments, and a certain amount of child like wonder, where someone was able to see things in a fresh new way. Everyone says they want to learn new things or to try something different. We are human beings after all, curious about the world around us. We want to probe, question, and experiment. And we are ecstatic about this process of exploration—until we realize someone is watching. A funny thing about people, nobody wants to look like a beginner. A beginner is clumsy and awkward and totally out of place and incredibly funny to watch—and to make fun of. A beginner is *The Newbie*. And who wants to be a *newbie?*

A similar anachronism is true about companies. Company managers will often say to their employees, "We want you to take risks. We want you to ask questions. We want you to think *outside of the box*." But the fact of the matter, in most companies someone is watching. And those who are watching usually do not have time for questions (especially dumb questions). They want answers, they want solutions—*and they want them now*. But how can you come up with a brilliant solution to a problem if you cannot ask any questions?

To properly identify the real problem you must be willing to engage in the process of exploration. It involves asking questions—lots of questions —followed by some serious thinking about the answers that follow. Sometimes the questions may seem dumb and some of the answers even dumber. Sometimes you might even need to have your answers questioned.

It is a simple process—but it is not easy. It requires some preparatory work, a lot of patience, a willingness to look a little bit silly, or clumsy and awkward, sometimes even downright stupid. But almost always the process produces a *damn good question*, or what I like to a call a *DGQ*. And asking that question—however simple it may seem—is what leads to success. Getting to that question takes time, effort, and a lot of hard work. It does not just pop-up in a moment of fantasy.

8.6.3 Determine the Origin

With parts made from thermoplastics, the real problem usually falls into one of four main categories: material issues, processing issues, tooling issues, or design issues. These categories are interrelated, but usually, a problem has an origin in one area (Figure 8.3).

We often start an investigation by looking for problems with the material. Indeed, "bad material" is often blamed for all kinds of problems. But the real issue here is whether the material, that is the raw material as it is coming into the factory, meets the specifications. Is it the material that was specified, in the proper grade and version? Or is it a substitute material, or a replacement material, or even a counterfeit material?

The key question here should be: *Is the vendor using the material that was specified?* (Is the vendor using the material that you carefully selected and specified after a comprehensive evaluation process?) The answer to this question is simple: it is either yes or no. If the material matches the specifications and meets the standards for that grade and version, the

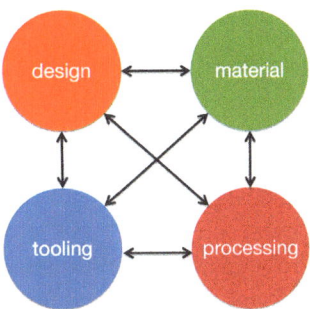

Figure 8.3 The four pillars of a plastic project.

answer is *Yes*. Therefore, it is not a "bad" material. (It may be the "wrong" material for the application, but that is a design issue, *not* a material issue.) If the answer is *No*, it means the vendor is using a material that was not specified. It could be a bad material, the wrong material, or even the right material that was used for the wrong reasons. But if a supplier is using a material that does not meet the specifications, this should be a deal breaker. *Tilt. Game over. Do not pass go. Do not move forward.*

The next category of issues is processing issues. Most often, we think of a processing issue as a problem that was caused by improper process control during injection molding but this category includes all of the processes that the material undergoes as it is transformed into a finished part. The key here is to confirm that the material in the finished part has the desired properties and performance.

Processing issues are typically the result of improper or insufficient process control. There are a myriad number of reasons for this, and there are hundreds if not thousands of books and technical papers on the thermoplastics processing. Establishing the proper process controls for a given application often takes some advanced expertise. However, it is important to remember that there may be a situation where we can identify that certain process controls are "wrong," but the real issue was that performance criteria for the final parts were not properly communicated in the first place. Next, there could be tooling issues. These kinds of issues are often interrelated with processing, and we may confuse them with a tooling issue. There could be an injection mold that is poorly designed, and/or poorly maintained. As a result, it is not capable of fabricating parts that meet the specifications. It may have a gate that is poorly sized, or in a bad location, or the sprue may be undersized, or the cooling lines are not optimized. The problems show up in the molded parts, so your investigation may focus on the processing of the material, but the reality is that the tool itself is the problem. Tooling issues are often hard to pinpoint, and when they occur, they are often hard to fix as well.

Finally, there are design issues. More often than not, thermoplastic parts fail because of bad design. But this category includes not only poorly designed parts, but errors in material selection as well. Selecting the "wrong" material is a design issue. Hopefully this book has provided enough guidance to help you select the "right" material.

Understanding these four categories of issues is an important aspect of plastics troubleshooting. Identifying the real problem, and tracing it back to its origin, is sometimes simple and straightforward. There are other times where the origin of the problem can be difficult to trace. Sometimes the problem may have an origin in multiple categories.

Years ago, I was asked to help troubleshoot a small problem with a hose mender. A hose mender is a simple little device, consisting of a short tube and a couple of tube clamps, which is used to repair a damaged section in a common garden hose. There are numerous varieties available; they might be made from steel, brass, plastic, or a combination of materials. This one was made from a standard grade of 6/6 nylon from DuPont. For some unknown reason, there had been an increase in the product failure rates, and the manufacturer had asked DuPont to help investigate.

The failure mode was pretty simple: the screw boss where the clamp was screwed together was cracking. It was not cracking on every part, but it was happening more often than before. After some investigation, we were able to determine that the failures were occurring most frequently in parts from one specific cavity in one specific mold, and most frequently in parts that were colored yellow. Further investigation revealed that there had been a change in the chemistry of the yellow colorant. It turned out that one of the ingredients in this colorant was also acting as a nucleating agent, which then caused a subtle variation in the degree of crystallinity in the molded parts and—more importantly—a subtle variation in the amount of mold shrinkage. The six sigma affect was that in certain mold cavities the screw holes were now just a touch too small—and these were the parts that were failing. The solution was to slightly increase the diameter of the hole—on all of the screw bosses in every cavity, in every tool. This was done by changing the core pins that formed these holes during the next round of mold maintenance. The funny thing is this was the only feature on the part with a critical dimension. While the tolerances were under control, the nominal dimension was just a little bit off. The change in colorant brought the issue to the forefront and helped guide our troubleshooting efforts (Figure 8.4).

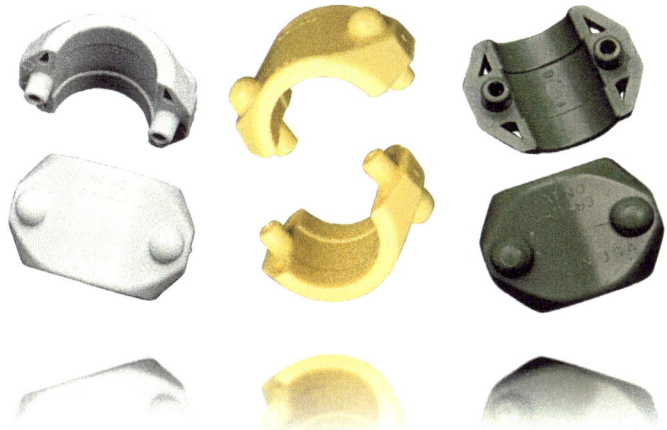

Figure 8.4 Hose mender parts, molded from 6/6 nylon.

8.6.4 Understand the Root Cause

Once you have identified the real problem, and traced it back to its origin, you can then look for the root cause, which is the factor that led to the problem occurring in the first place.

There is a tendency with many companies to focus on the origin of the problem, rather than understanding the root cause behind it. As an example, let us assume a part is breaking, and we determine that the material that was selected did not have the needed properties. This is a design error, and it is a common one. You could say the error was due to bad judgment or a failure to do due diligence. However, the root cause of that error was more likely to be a lack of training, or a lack of expertise in thermoplastic materials, or even with the selection process itself (assuming there was an actual process). Unless changes are made to address the root cause, it is likely that the design team will make errors in material selection again in the future.

In a similar manner, there are often problems that can be traced to equipment failures or to operator error. It is common to focus on the mistakes that were made. Were they errors in judgment? Or were they due to a lack of experience or even ignorance? Regardless of the error itself, the root cause may have been one of missing procedures, or improper procedures. This is particularly important for equipment maintenance. If the proper procedures are not put in place, problems will reoccur.

Sometimes, the root cause is a lack of skill, or that a basic capability is missing. These causes are often difficult to accept, as they can reveal serious flaws in a company culture or in its processes. But no matter what the problems are, or what their root causes are, the solution to the problem needs to address the root cause. Some people might ask, "Why would a company try to solve a problem without addressing the underlying issues that caused the problem in the first place?"

Now that's a damn good question.

8.6.5 Solve the Problem(s)

Once you have identified the real problem, tracked it to its origin, and found the root cause, finding a solution is usually pretty simple. The bigger challenge is often in getting everyone to agree with the findings of the troubleshooting team. Much like a material specification document, a troubleshooting report should be clear and concise with a specific action plan.

Once the solution is implemented, it is often helpful to review the troubleshooting process itself, either in a *Lessons Learned* review, or as part of a *Continuous Improvement* program. That review may also involve a reassessment of your vendor-sourcing protocol, not just in whether a given vendor is actually capable, but in the methods that were used to evaluate that vendor in the first place.

On occasion, the solution may involve changing vendors. This is a drastic solution and often involves the biggest disruption to the supply chain. There will be a learning curve with the new vendor, involving not only new materials and/or parts, but new relationships, new protocols, and new communication efforts. While this new vendor may provide the solution that is needed, the transition needs to be carefully managed.

Troubleshooting is often a thankless task. While it is normally quite obvious that there is a problem, the fact finding mission alone can be arduous. You are asking questions and identifying issues, and if it is not done delicately, there may be those who feel you are stepping on toes. You may be uncovering problems with technology and logistics, or missed communication, poor handoffs, and dropped batons, to say nothing of the politics of the business including turf wars and political minefields. The best troubleshooters are aware of all this, and go about the job with tact and professionalism (Figure 8.5).

8.7 Finale

Throughout history, humans have always made things. This includes items that were needed for basic survival—things like tools, weapons, clothing, and shelter—as well as items that were intended to help fulfill higher needs—things like games, toys, art, music, and even jewelry.

8: Bringing It All Together

Figure 8.5 Don't throw the engineers under the bus.

The materials that humans have used to make things have included naturally occurring materials such as wood and stone, as well as found materials such as shells and antlers and bones and feathers. In more recent history, humans have taken naturally occurring materials such as aluminum and copper and iron and—after refining them to a pure state—they have used these refined materials to make items that meet a wide variety of needs. Today, many of the things that humans make are made from synthetic materials, including concrete, steel, and plastics.

Regardless of the item being made, the selection of the right material has always been challenging task. This is true not only for the craftsman and artisan who are making unique one-of-a-kind items, but for the designers and engineers of items that will be used *en masse*.

In this book, we have focused our efforts on the effective use of thermoplastics. Our intent has been to provide the reader with the tools they need to make informed decisions about these materials.

We hope you have found this information helpful.

There is no I in good design.
                               Eric R. Larson

Resources

Plastics Publications/Web Sites

Plastics Business

Plastics Business is a quarterly magazine targeting the plastics processing business executive. It is distributed to corporate management as well as plant and production managers involved with all types of plastics processing and manufacturing, primarily in the United States.

www.plasticsbusinessmag.com

Plastics Engineering Magazine

Plastics Engineering magazine is published on behalf of the Society of Plastics Engineers by John Wiley & Sons Inc.

www.plasticsengineering.org

Plastics News

Plastics News and its sister publications together comprise the world's largest plastics publishing organization. Owned by Crain Communications Inc., these brands include *Plastics News*, *Plastics News China*, *European Plastics News* and *Plastics & Rubber Weekly.* Other polymer-related publications owned by Crain include *Rubber & Plastics News, European Rubber Journal, Urethanes Technology International,* and *Tire Business.*

www.plasticsnews.com

Plastics Technology

Monthly print magazine and online newsletter published by Gardner Business Media, Inc. Related publications owned by Gardner include *Automotive Design & Production, CompositesWorld, Modern Machine Shop, MoldMaking Technology, Production Machining, Products Finishing.*

www.ptonline.com

plastics.com

plastics.com is the world's largest online plastics community with over 70,000 members and features technical advice, blogs, calendars, resources, articles, news, forums, and a marketplace.

www.plastics.com

plasticsguy.com

A blog site with information on the art and science of mass production, including design, materials (especially plastics), and processing, as well as musings on the business of design.

www.plasticsguy.com

Professional Societies

American Chemical Society (ACS)

An independent membership organization which represents professionals at all degree levels and in all fields of chemistry and sciences that involve chemistry. Has Technical Divisions in Polymer Chemistry, and Polymeric Materials: Science & Engineering.

www.acs.org

American Society of Mechanical Engineers (ASME)

ASME is a not-for-profit membership organization that enables collaboration, knowledge sharing, career enrichment, and skills development across all engineering disciplines, toward a goal of helping the global engineering community develop solutions to benefit lives and livelihoods. Has occasional papers on design with plastics.

www.asme.org

SAE International

SAE International is a global body of scientists, engineers, and practitioners that advances self-propelled vehicle and system knowledge in a neutral forum for the benefit of society. Offers standards, books, technical papers, and training.

topics.sae.org/polymers

Society for the Advancement of Material and Process Engineering (SAMPE)

An international professional member society, provides information on new materials and processing technologies through chapter technical presentations, two journal publications, symposia, and commercial expositions in which professionals can exchange ideas and air their views.

www.nasampe.org

Society of Plastics Engineers (SPE)

SPE is the largest, most well-known plastics professional society in the world.

www.4spe.org

Society of Plastics Engineers—Thermoforming Division

Technical Division to facilitate the advancement of thermoforming technologies through education, application, promotion, and research.

www.thermoformingdivision.com

Society of Plastics Engineers—Blow Molding Division

Technical Division interested in all aspects of blow molding: ranging from machinery, tooling and auxiliary equipment, to polymer properties, processing, and fabricated part testing.

www.blowmoldingdivision.org

Standards Organizations

American National Standards Institute (ANSI)

A private nonprofit organization that oversees the development of voluntary consensus standards for products, services, processes, systems, and personnel in the United States.

www.ansi.org

ASTM International (ASTM)

Previously known as the American Society for Testing and Materials, ASTM is an international standards organization that develops and publishes voluntary consensus technical standards for a wide range of materials, products, systems, and services.

www.astm.org

International Organization for Standardization (ISO)

An international standard-setting body composed of representatives from various national standards organizations.

www.iso.org

Underwrlters Laboratories (UL)

A safety consulting and certification company headquartered in Northbrook, Illinois. Primarily focused on drafting of safety standards for electrical devices and components, but also offers standards for plastics materials, including flammability.

www.ul.com

Trade Organizations

Society of the Plastics Industry (SPI)

SPI promotes growth in the US plastics industry.
www.plasticsindustry.org

PlasticsEurope

PlasticsEurope is one of the leading European trade associations with centers in Brussels, Frankfurt, London, Madrid, Milan, and Paris.

www.plasticseurope.org

American Mold Builders Association (AMBA)

A leading association of mold builders representing a large variety of processes and capabilities.

www.amba.org

Association of Rotational Molders

A worldwide trade association currently representing member companies in 58 countries. Members include manufacturers of rotationally molded plastic products, suppliers to the industry, designers, and professionals.

www.rotomolding.org

Trade Shows

K Trade Fair

Known as the K show (K as in *Kunststoff*, the German word for plastic), it has been and still is the biggest international trade fair for plastics and rubber, and takes place once every 3 years. The next show is in 2016 in Dusseldorf, Germany.

www.k-online.com

National Plastics Expo (NPE)

The world's leading plastics trade show and conference, NPE is held every 3 years: 2015, 2018, 2021, etc. The 2015 conference will be held in Orlando, FL.

www.npe.org

Pacific Design & Manufacturing Show

Held in Anaheim, CA every February, the Pacific Design Show is one of the largest engineering trade shows in the western US. It is held in conjunction with MD&M West, ATX West, WestPack, Electronics West, AEROCON, Quality Expo, PLASTEC West.

pacdesignshow.designnews.com

Academia

Auburn University

Auburn, AL
Samuel Ginn College of Engineering
Polymer and Fiber Engineering Program
BS, MS, and PhD degrees
eng.auburn.edu/programs/pfen/index.html

Ferris State University

Big Rapids, MI
BS degree in plastics engineering technology
www.ferris.edu/plastics-engineering-degree.htm

Lehigh University

Bethlehem, PA
Materials Science and Engineering Department
Center for Polymer Science and Engineering
MS, ME, and PhD degrees
www.lehigh.edu/~inpolctr/index.html

Pennsylvania College of Technology

Williamsport, PA
Associates and Bachelor degrees in Polymer Engineering and Technology
www.ptc.edu/schools/icet/plastics

Pennsylvania State University—Erie-Behrend College

Erie, PA
BS degree in plastics engineering technology
psbehrend.psu.edu/school-of-engineering/academic-programs/plastics-engineering-technology

Stevens Institute of Technology

Hoboken, NJ
Masters degree in polymer chemistry and polymer engineering
www.stevens.edu

University of Akron

Akron, OH
College of Polymer Science and Polymer Engineering
BS, MS, and PhD degrees
www.uakron.edu/cpspe/

University of Massachusetts Amherst

Amherst, MA
Department of Polymer Science and Engineering
Offers graduate programs only
www.pse.umass.edu/about-pse

University of Massachusetts Lowell

Lowell, MA
Plastics Engineering Department
BS, MS, and PhD degrees
www.uml.edu/engineering/plastics

University of Southern Mississippi

Hattiesburg, MS
School of Polymers and High Performance Materials
BS, MS, and PhD degrees
www.usm.edu/polymer

University of Tennessee

Knoxville, TN
MS and PhD degrees in Polymer Chemistry and Polymer Engineering
http://www.engr.utk.edu

University of Wisconsin

Madison, WI
Polymer Engineering Center
MS and PhD degrees in Polymer Engineering and Polymer Science
pec.engr.wisc.edu/education.html

Index

Note: Page numbers followed by "f" and "t" indicate figures and tables respectively.

A
Acetal, 109–111
  copolymer, 110
  disposable lighters molded from, 111, 112f
  homopolymer, 110–111
Acrylic, 98, 106t–107t. *See also* Polymethyl methacrylate (PMMA)
Acrylonitrile butadiene styrene (ABS), 108–109
Age of Metals, the, 3–7
  bronze object, 4, 5f
  copper ore, 3, 4f
  iron ore, 4–5, 6f
Alloys, 86–87
Amorphous, 66
  versus semicrystalline, 92–93
Amortization, 213

B
Benchmarks, 146
Bending, 58, 59f, 60
  and stiffness, 61–62, 61f–62f, 166
Blends, 86–87
Blow molding, 37
  extrusion blow molding, 37
  injection blow molding, 38
  for plastic bottle, 39f
  preforms, 38f
Boat anchor (slang), 214

C
Cantilevered beam with end load, 153, 153f
Capital, 213
Capital asset, 213
Capital investment, 213
Cellulose, 13
Cellulose acetate, 271
  hair barrette made from, 271, 271f
Ceramics, 7–8

Charpy test, 171, 172f
Clays, 7–8
Commodity plastics, 105, 106t–107t
  acrylic, 98
  ethylene propylene diene monomer (EPDM), 102
  polyacrylates, 98
  polyethylene (PE). *See* Polyethylene (PE)
  polymethyl methacrylate (PMMA), 98–100, 116
  polypropylene (PP). *See* Polypropylene (PP)
  polystyrene (PS), 102–103
  polyvinyl chloride (PVC), 103–104
  thermoplastic polyurethane (TPU), 105
Compression, 58, 59f, 165, 165f
Compression molding, 31
  antique waffle iron, 31, 31f
Conductive plastics, 191
Copolymer, 64–66, 110
Correlation
  correlation model, 148, 148f
  in predicting performance, 148–149
Cost of goods, 213
Costs, 207–209, 208f
  evaluating, 214–219
    adding up numbers, 219
    cycle times, 217–219
    material cost, 215–216
    process rates, 216–217, 218t
    processing cost, 216
  importance of, 209–212
    bottom line on cost, 211–212, 211f
    business perspective, 210–211
    measuring cost, 210
    relationship to performance, 210
  language of, 212–214
  reduction. *See* Material costs, reducing; Processing cost, reducing
  in thermoplastic material selection, 208

339

Cracking, 151
  initiation and propagation, 168
Critical material properties, 203
Cross-linking process, 23–24

D
Design, 214
  D F X (design for initiative), 230–231
  importance of, 152–154
  options, and manufacturing costs, 243
  for speed, 241–242
Drop testing, 174–175, 175f

E
Elastic deformation, 58–59
Elastic modulus, 60–61
Electrical impulses, in sensing
  sound, 273, 275
  smell, 293–294
  taste, 300
  touch, 282
Electrical properties, 191
  conductive plastics, 191
  insulating plastics, 191
Electromagnetic waves, 160
  spectrum, 161f
Engineering plastics, 108–119, 119t–120t
  acetal, 109–111. *See also* Acetal
  acrylonitrile butadiene styrene (ABS), 108–109
  polyamide. *See* Polyamide (PA)
  polyester, 116–118
  polyethylene terephthalate, 116–117
  polyphenylene oxide (PPO), 118–119
Ethylene propylene diene monomer (EPDM), 102, 137
Ethylene tetrafluoroethylene (ETFE), 125–126
Expandable foam molding, 36
  coffee cups, 36, 36f
Extrusion, 29–30
  extruded drinking straws, 29f
Extrusion blow molding, 37

F
Failure mode and effects analysis (FMEA), 199
Feel, 251–253
  evaluating, 265–266
    human senses, 266. *See also* Human senses, in evaluating feel
    physical equipment, 266
  importance of, 253–256
    product performance, 253–254
  language of, 256–265, 258f
    comparative analysis, 265, 265f
    human response, 259–263, 260f, 262f
    imprecise language, 264
    non-engineering, 263–264, 264f
    sensory input, 257–259
  sales and market share, 254–255, 255f
    bottom line on feel, 256
    technical validity, 255–256
  thermoplastic material selection, methodology, 302–307
    flow chart based on feel, 306, 307f
    infrastructure, 303–304
    process, 304–306
    suggestions, 306–308
  touch, 251–252, 252f–253f
Flammability, 160
Flexural modulus, 61, 61f, 166–167
Flexural strength, 61, 61f
Fluorinated ethylene propylene (FEP), 126, 134t–135t
Fluoropolymers, 124–125, 134t–135t. *See also* Fluorinated ethylene propylene (FEP); Perfluoroalkoxy alkane (PFA)
Foam molding
  expandable, 36
  structural, 35–36
Fordism, 10
Form properties, 192–197
  appearance, 195–197
    material samples, standard colors of, 196–197, 197f
  shape, 194–195
  size, 192–194
    Herman Miller Equa chair shell, 192, 193f
    micromolding, 194, 194f
Fracture mechanics, 168

Index

G
Gamma rays, 161–162
Gardner impact testing, 172–173
Glass, 7–8
Glass transition temperature, 67–68

H
Hardness, 289–291
　data comparison, 290, 290f
Heat deflection temperature
　　(HDT), 156, 199
High-speed tensile tests, 173–174
Homopolymer, 64–66, 110–111
Housings, in Apple products, 45–46
Human senses, in evaluating feel, 266
　hearing, 273–279
　　acoustics, 274–275
　　human response, 277
　　material selection based on,
　　　277–279, 279f
　　music, 276–277
　　opportunities, 279
　　psychoacoustics, 275–276
　　sheet music, 276, 277f
　　sound, 273–274
　　sound waves, 273–274, 274f
　　vibrations, 274
　sight, 266–273
　　color, 269–270, 270f
　　light, 267–269
　　material selection based on,
　　　271–273, 272f
　　opportunities, 273
　　optical data, 267, 268t
　　patterns, 270–271, 271f
　smell, 293–299, 294f
　　human brain, 294–295, 295f
　　human response to odor, 298
　　material selection based on, 299
　　new car smell, 297, 297f
　　odor detection, 294–296
　　odor in thermoplastics, 296–298
　　opportunities, 299
　taste, 300–302
　　material selection based on, 301–302
　　opportunities, 302
　touch, 279–293
　　hardness, 289–291
　　material selection based on, 292–293

　　movement, 287–289, 289f
　　opportunities, 293
　　pressure, 286
　　size and shape, 280
　　slipperiness, 291–292, 292f
　　temperature, 282–286
　　texture, 291
　　vibration, 286–287
　　weight and density, 280–282,
　　　281f–282f
Hydroslide, 180

I
Industrial Revolution, 8–10
　Aubin forging mills, 8–9, 9f
Industry infrastructure,
　　317–320
　education, 318–319
　information providers, 319
　plastic testing, 319
　service providers, 319–320
　trade organizations, 318, 318f
Injection blow molding, 38
Injection molding, 32–33, 213
　different screws for, 32–33, 33f
　reaction injection molding, 34
Instrumented impact tests, 173
Insulating plastics, 191
Izod test, 170
　schematic diagram of pendulum test,
　　170, 171f

J
Job production, plastics
　　manufacturing, 44

K
Kevlar® (aramid fiber), 122–123

L
Leaf blower, 181, 181f
Liquid crystal polymer (LCP),
　　126–127
　multipin connectors molded
　　from, 127f

M
Mass production, 10–11
Material cost, 213

Material costs, reducing, 219–230
  effective specifications, 222–224
  exploiting competitive advantages, 227–229
    Ethafoam®, 229
    Mach 7 product line, 229, 230f
    SRAM IBS shifters, 227–228, 227f
    Surlyn®, 229
  exploiting material, 224–226
    BIC lighters, 225, 225f
    Fresnel lens, 226, 226f
  optimizing structure, 220–222
    five-legged office chairs, 220, 221f
  optimizing wall thickness, 224
  reducing processing cost, 230–242. *See also* Processing cost, reducing
Material properties effects analysis (MPEA), 200, 201t–202t
Material selection, 311–313
  based on
    hearing, 277–279, 279f
    sight, 271–273, 272f
    smell, 264
    taste, 301–302
    touch, 292–293
  data, evaluating, 312–313
  development, 313
  environmental effects, 156–164
    chemicals, 159–160
    four horsemen of plastic apocalypse, 157, 157f
    radiation, 160–162
    temperature, 158–159
    time, 162–164
  infrastructure. *See* Industry infrastructure
  key criteria, establishing, 312
  manufacturing process, selecting, 312
  material candidates, list developing, 312
  material, selecting, 313
  opportunities
    hearing, 279
    sight, 273
    smell, 264
    taste, 302
    touch, 293

  and performance, 150–156, 151f
    design, importance of, 152–154, 153f
    processing, importance of, 154–155
    property data. *See* Property data
  specification of. *See* Material specification
  supply. *See* Supply chain, plastics
Material selection and cost, 207–250
  costs, 207–209, 208f. *See also* Costs
    evaluating, 214–219
    importance of, 209–212
    language of, 212–214
    reducing costs. *See* Material costs, reducing; Processing cost, reducing
    in thermoplastic material selection, 208
  total manufacturing cost, 243–247. *See also* Total manufacturing cost
Material specification, 313–315
  approved suppliers, 314–315
Materials science, 3, 11–12, 58–64
  anisotropic behavior, 60–61
  nonlinear behavior, 62
  professional societies, 12
  role of chemistry in, 8
  stiffness, 62–63, 62f
  strength of materials, 58–60
    stress–strain curves, 59, 59f
    types of load, 58, 59f
  toughness, 63–64, 63f. *See also* Toughness
Mechanical properties, 164–170
  stiffness, 166–167
  strength, 164–166
    types of loads, 165, 165f
  toughness, 167–170. *See also* Toughness
Melt cycle, 69
Melt temperature, 68
Melting temperature, 68
Model T, 10
  factory, 11f
Molding, 25
  blow molding. *See* Blow molding
  closed mold, 26f
  compression molding, 31
  foam molding. *See* Foam molding
  injection molding. *See* Injection molding

INDEX 343

open mold, 25f
plastic molding, 30
rotational molding, 32
transfer molding, 34–35
Molecular weight, 68
 distribution, 68
Moment of inertia, 153
Monomers, 64
 propylene monomer, 65f

N

Nomex® (aramid fiber), 122
Noryl®, 119
Notch sensitivity, 168
 window glass, 168f
Nylon (polyamide), 13, 112, 184
 amorphous nylon, 114
 nylon 11, 114
 nylon 12, 114
 nylon 6, 113
 nylon 6/10, 113
 nylon 6/12, 113
 nylon 6/6, 113
  effects of oven aging on, 162, 163f
  hose mender parts, molded from, 326, 327f
 nylon chemistry, 114
 parachute, 14f

O

Odor
 human response to, 298
 in thermoplastics, 296–298
Off-spec, 77, 77f
Optical data, 267, 268t
Optical grade, 77
Optimization
 for reducing material costs
  of structure, 220–222
  of wall thickness, 224
 for reducing processing costs, of geometry, 231–233

P

Paleoanthropology, 1–2
PBS NOVA program, 5–6
Perfluoroalkoxy alkane (PFA), 126
Performance, 145–147, 147f
 and material selection, 150–156, 151f.
  See also Material selection

design, importance of, 152–154, 153f
processing, importance of, 154–155
property data. *See* Property data
predicting, 147–150
 correlation, 148–149
 disruptive innovation, 149–150
 wrong criteria, 149
Phono preamplifier with tubes, 278, 279f
Physiology, context of sound, 275
Plastic Age, the, 12–14, 17
Plastic behavior, 19, 19f
Plastic deformation, 19, 58–59
Plastic processing technologies, 24–38
 blow molding, 37
 casting, 30–31
 compression molding, 31
 expandable foam molding, 36
 extrusion, 29–30
 extrusion blow molding, 37
 injection blow molding, 38
 injection molding, 32–33
 plastic molding, 30
 plastic welding, 28–29
 plastics forming, 26
 plastics tooling, 25
 pressure forming, 27–28
 process comparison, 39
  common processing techniques, 39, 40t–41t
 reaction injection molding, 34
 rotational molding, 32
 structural foam molding, 35–36
 transfer molding, 34–35
 vacuum forming, 26–27
Plastics, 1, 12, 14–15, 20
 in manufacturing, 42–46
  batch production, 43–44
  job production, 44
  manufacturing processes, 44–46
  mass production, 42–43
 market use, 70, 71f
 material selection, 15–16
 processing technologies, 24–38. *See also* Plastic processing technologies
 as raw materials, 20–24
 thermoplastics. *See* Thermoplastics
 thermosets. *See* Thermosets

Plastics, academia
  Auburn University, 335
  Ferris State University, 336
  Lehigh University, 336
  Pennsylvania College of Technology, 336
    Erie-Behrend College, 336
  Stevens Institute of Technology, 336
  University of Akron, 336
  University of Massachusetts Amherst, 337
  University of Massachusetts Lowell, 337
  University of Southern Mississippi, 337
  University of Tennessee, 337
  University of Wisconsin, 337
Plastics publications
  Plastics Business (magazine), 331
  Plastics Engineering (magazine), 331
  Plastics News, 331
  Plastics Technology (magazine), 331
Plastics technology, 13–14
Plastics tooling, 25
Plastics websites
  plastics.com, 332
  plasticsguy.com, 332
Polyacrylates, 98
Polyamide (PA), 111, 114
  cable tie, 115f
  polyamide 11, 114
  polyamide 12, 114
  polyamide 6, 113
  polyamide 6/10, 113
  polyamide 6/12, 113
  polyamide 6/6, 113
Polyamide-imide (PAI), 129
Polyarylate (PAR), 127
Polybutylene terephthalate (PBT), 116–118
  hot melt glue gun, 118, 118f
Polycarbonate, 115, 184
  football helmet, 148–149
  modern fighter jets, 116f
Polycyclohexylenedimethylene terephthalate, 117
Polyester block copolymers (PBC), 138–139
  constant-velocity (CV) joint, 139, 139f
Polyester, 116–118
Polyether-block-amide elastomer, 140

Polyetherimide (PEI), 129
Polyethersulfone, 133
Polyethylene (PE), 100–102
  bubble wrap packaging, 101, 101f
  polymers, 100–101
Polyethylene terephthalate (PET), 116–118
  hot melt glue gun, 118, 118f
Polyethylene terephthalate, 116–117
Polyimide (PI), 128–129
  flexible printed circuits, 128f
Polyketone (PK), 130
Polymer molecule, 20, 20f
Polymer science, 64–69
  amorphous, 66
  crystallinity, 66
    molecular model of, 67f
  crystallization, 66–67
  glass transition temperature, 67–68
  melt cycle, 69
  melt temperature, 68
  melting temperature, 68
  molecular weight, 68
    distribution, 68
  polymer chemistry, language, 65f, 66
  resin, 66
Polymerization, 64
Polymers, 12–13
  and monomers, 64
  polypropylene polymer, 65f
Polymethyl methacrylate (PMMA), 98–100, 116
  in aquariums, 100, 100f
Polymethylpentene (PMP), 130
  laboratory glassware, 131f
Polyolefin blend elastomers (POE), 137–138
Polyphenylene oxide (PPO), 118–119
Polyphenylene sulfide (PPS), 130–131
Polyphenylsulfone, 133
Polyphthalamide (PPA), 131–132, 132f
Polypropylene (PP), 102
  copolymer, 102
    block copolymer, 102
    impact-modified copolymer, 102
    random copolymer, 102
    recycling bin molded from, 102, 103f
  homopolymer, 102
  monomer of, 123f

INDEX

Polystyrene (PS), 102–103
   CD cases made from, 103, 104f
Polysulfone (PSU), 132–133
Polytetrafluoroethylene (PTFE), 124–125
   thread sealing tape made from, 125, 125f
Polytrimethylene terephthalate (PTT), 117
Polyvinyl chloride (PVC), 103–104
   vinyl LP records, 104f
Polyvinylidene fluoride (PVDF), 126
Pressure forming, 27–28
   used in conjunction with vacuum, 27, 28f
Processing cost, 213
Processing cost, reducing, 230–242
   design for speed, 241–242
   effective specifications, 233–237
      parts and assembly, 237, 238f
      precision and price, 234, 235f
   exploiting materials, 239–241
   optimizing geometry, 231–233
   uniform wall thickness, 238–239
      chair base design, 239, 240f
Professional societies
   American Chemical Society
      (ACS), 332
   American Society of Mechanical
      Engineers (ASME), 332
   SAE International, 332
   Society for the Advancement of
      Material and Process Engineering
      (SAMPE), 333
   Society of Plastics Engineers
      (SPE), 333
      Blow Molding Division, 333
      Thermoforming Division, 333
Projectile testing, 174
Property data, 155–156
   evaluation, 152
Psychology, and perception, 276

Q

Qualitative analysis, 176
Quantitative analysis, 199–200

R

Radiofrequency (RF) spectrum, 161
Rates
   cost rates, 217, 218t
   process rates, 216–217

Reaction injection molding (RIM), 34
Resin, 66
Resin industry, 69–90
   alloys and blends, 86–87
      oil and vinegar blend, 86, 87f
   fillers, 82
   performance modifiers, 83–84
   processing aids, 81–82
   reinforcements, 82–83
      glass fiber "chopped strands", 83, 83f
   resin distribution, 71–75
      bag, 72, 73f
      boat load, 74
      bulk box, 72, 73f
      railcar, 72–74, 74f
      truckload, 72
   resin grades, 75–79
      food grade, 75, 76f
      generic prime, 75, 76f
      industrial, 77
      medical grade, 76
      off-spec, 77, 77f
      optical grade, 77
      postconsumer waste, 78
      preconsumer waste, 78
      prime, 75
      recycled, 77–78, 78f
      regrind, 78
      reprocessed, 77
      virgin, 78, 79f
   resin modification, 79–81
      compounding, 80
      cube blending, 80, 81f
      dry blending, 80–81
      melt blending, 80
   resin production, 70–71
   resin versions, 84–86
Risk priority number (RPN), 199
Rotational molding, 32
   molded parts, 32, 32f

S

Semiotics, 276
Sensation, touch, 251–252, 252f–253f,
      279–293
   hardness, 289–291
   material selection based on, 292–293
   movement, 287–289, 289f

Semiotics (*Continued*)
  opportunities, 293
  pressure, 286
  size and shape, 280
  slipperiness, 291–292, 292f
  temperature, 282–286
  texture, 291
  vibration, 286–287
  weight and density, 280–282, 281f–282f
Shear, 58, 59f, 165, 165f
Smell, 293–299, 294f
  material selection based on, 299
  odor
    detection, 294–296
    human brain, 294–295, 295f
    human response to, 298
    new car smell, 297, 297f
    in thermoplastics, 296–298
    opportunities, 299
      aroma of freshly molded plastics, 299, 299f
Sound, 273–274
  sociology, 276
  sound waves, 273–274, 274f
    reception of, 273–275
    transmitted by tuning fork, 287f
Specialty plastics, 134t–135t, 122–133, 134t–135t
  aramid, 122
    speaker cone fabricated from, 124f
  fluorinated ethylene propylene (FEP), 126
  fluoropolymers, 124–125
  liquid crystal polymer (LCP), 126–127
  perfluoroalkoxy alkane (PFA), 126
  polyamide-imide (PAI), 129
  polyarylate (PAR), 127
  polyetherimide (PEI), 129
  polyethersulfone, 133
  polyimide (PI), 128–129
  polymethylpentene (PMP), 130
  polyphenylene sulfide (PPS), 130–131
  polyphenylsulfone, 133
  polyphthalamide (PPA), 131–132, 132f
  polysulfone (PSU), 132–133
  polytetrafluoroethylene (PTFE), 124–125
  polyvinylidene fluoride (PVDF), 126
  ultrahigh molecular weight polyethylene (UHMWPE), 133
Specification, 223
  effective, 233–237. *See also* Material specification
Sporting goods, 185–186
SRAM (manufacturer), 185–186
  Shupe test, 187–188
Standards organizations
  American National Standards Institute (ANSI), 333
  ASTM International (ASTM), 334
  International Organization for Standardization (ISO), 334
  Underwriters Laboratories (UL), 334
Stiffness, 62–63, 153, 166–167
  versus weight, 47
  to withstand bending, 23f, 25f, 61–62
Stiffness factor, 153–154
Stone Age, the, 1–3
  sample of early stone tools, 1–2, 2f
Structural foam molding, 35–36
  trash receptacle, 35, 35f
Stryofoam™, 103
Styrene
  ABS, 108–109
  monomer of, 123f
Styrene acrylonitrile (SAN), 104–105
Styrenic block copolymers (SBC), 136
Suppliers
  approved, 314–315
  working with, 320–322
    communication, 322
    determining capabilities, 320–321
    determining right fit, 321
    managing relationship, 322
    project participation, 321
Supply chain, plastics, 315–317, 315f
  compounders, 316
  converters, 316
  equipment suppliers, 316
  product manufacturers, 317
  resin suppliers, 316
  toolmakers, 317
Surface properties, 188–190
  friction, 189
  hardness, 190

INDEX 347

lubricity, 189–190
wear, 190
Synthetic materials, 13–15
man-made materials, 55

T

Taste, 300–302
　material selection based on, 301–302
　　taste of plastic, 301, 301f
　　opportunities, 302
　　edible plastics, 302, 302f
Tear resistance, 168, 169f
Teflon®, 1, 101
Tension, 58, 59f, 165, 165f
Terpolymer, 64–66
Thermal conductivity coefficient (TCC), 282, 284–286
　of metal horseshoe, 285, 285f
　of plastics and other materials, 283t–284t
Thermoforming, 126–127, 139, 155
Thermoplastic classification methods, 90–94
　amorphous versus semicrystalline, 92–93
　chemical family, 93
　cost versus performance, 93–94
　elasticity, 94
　tree of life, 91f
Thermoplastic elastomers (TPEs), 97–98, 135–140
　collage of items made from, 135f
　elastomeric alloys, 138
　iDive housing, 137, 137f
　polyamide elastomers, 139
　polyester block copolymers (PBC), 138–139
　polyether-block-amide elastomer, 140
　polyolefin blend elastomers (POE), 137–138
　styrenic block copolymers (SBC), 136
Thermoplastic materials, 97–133, 311–312
　commodity plastics, 98–105
　　examples of products, 97–98, 99f
　engineering plastics, 108–119. *See also* Engineering plastics

industry infrastructure. *See* Industry infrastructure
material selection. *See* Material selection
price–performance–volume chart, 97, 98f
specialty plastics, 122–133. *See also* Specialty plastics
troubleshooting, 322–328
　origin, determination, 324–326
　problem solving, 328, 329f
　real problem identification, 323–324
　team assembling, 323
　understanding root cause, 327–328
Thermoplastic materials, guidelines, 197–203
　critical material properties, 203
　manufacturing team, 198
　mathematical tools, 199–200
Thermoplastic polyurethane (TPU), 105
　automotive dashboards, 105
Thermoplastic vulcanizates (TPV). *See* Polyolefin blend elastomers (POE)
Thermoplastics, 21–22
　advantages of, 46–49
　　cost, 48–49
　　near-net-shape manufacturing, 47
　　performance, 46–47
　　processing options, 47
　　safety, 48
　disadvantages, 49–55
　　heat resistance, 49
　　human behavior, 54–55
　　perception, 53–54
　　repairing broken parts, 52, 53f
　　structural inconsistencies, 51–52
　　temperature variations, 51
　　time-dependent behavior, 49–50, 50f
　and thermosets, 22–24
　uniqueness of, 55
Thermosets, 21
　common examples of, 23, 23f
　and thermoplastics, 22–24
3D printing, 46
Timber, 7
Tooling, 213
Tooling maintenance, 214
Tooling ownership, 214

Torsion, 58, 59f
Total manufacturing cost, 243–249
    calculating, 243–244, 245t–246t
        flow chart to, 247, 248f
    explore design options, 243
    inspiration in products, 248–249, 249f
    math of, 243–247
Toughness, 167–170
    bottom line on, 188, 189f
    chew toys, 168, 169f
    crack initiation and propagation, 168
    fracture mechanics, 168
    measuring, 170–176
        Charpy test, 171, 172f
        comparing tests, 177t–178t
        drop testing, 174–175, 175f
        Gardner impact testing, 172–173
        high-speed tensile tests, 173–174
        instrumented impact tests, 173
        Izod test, 170
        projectile testing, 174
        tumble testing, 175–176
        un-notched Izod test, 171
        velocity comparison, 179t
    notch sensitivity, 168
        window glass, 168f
    sudden impact, 167, 167f
    tear resistance, 168
Trade organizations, 318
    American Mold Builders Association (AMBA), 334
    Association of Rotational Molders, 335
    in plastic industry, 318f
    PlasticsEurope, 334
    Society of the Plastics Industry (SPI), 334
Trade shows
    K Trade Fair, 335
    National Plastics Expo (NPE), 335
    Pacific Design & Manufacturing Show, 335

Transfer molding, 34–35
Troubleshooting, 322–328
    origin, determination, 324–326
        plastic project, pillars of, 324, 325f
    problem solving, 328, 329f
    real problem identification, 323–324
    team assembling, 323
    understanding root cause, 327–328
Tumble testing, 175–176

U
Ulfberht swords, 5–6, 6f
Ultrahigh molecular weight polyethylene (UHMWPE), 133
Un-notched Izod test, 171
UV light, 160–162

V
Vacuum forming, 26–27
    blister pack, 26–27, 27f
Valve amplification, 261
    assortment of vacuum tubes, 261f
Vibration, 286–287
    transmitted by tuning fork, 287f

W
Water (H_2O), 159–160
Wood, 7

X
X-rays, 160

Y
Yield rate, 243–244
Young's modulus, 60

Z
Zippers, high-performance, 184

Edwards Brothers Malloy
Ann Arbor MI. USA
June 8, 2015